Emissions Trading as a Policy Instrument

CESifo Seminar Series
edited by Hans-Werner Sinn

See http://mitpress.mit.edu for a complete list of the series.

Emissions Trading as a Policy Instrument

Evaluation and Prospects

edited by Marc Gronwald and Beat Hintermann

 Seminar Series

The MIT Press
Cambridge, Massachusetts
London, England

MIT Press books may be purchased at special quantity discounts for business or sales promotional use. For information, email special_sales@mitpress.mit.edu.

Set in Palatino by Toppan Best-set Premedia Limited, Hong Kong. Printed and bound in the United States of America.

Library of Congress Cataloging-in-Publication Data

Emissions trading systems as a policy instrument : evaluation and prospects / edited by Marc Gronwald and Beat Hintermann.
    pages cm. – (CESifo seminar series)
Includes bibliographical references and index.
ISBN 978-0-262-02928-5 (hardcover : alk. paper)
    1. Emissions trading–European Union countries. 2. Environmental policy–Economic aspects–European Union countries. I. Gronwald, Marc, 1975– II. Hintermann, Beat, 1974–
HC240.9.P55E446   2015
363.738'7–dc23
                              2015001162

10   9   8   7   6   5   4   3   2   1

Contents

Series Foreword

The CESifo Seminar Series covers topical policy issues in economics, largely from a European perspective. The books in the series originate in seminars hosted by CESifo, an international research network of economists organized jointly by the Center for Economic Studies at Ludwig-Maximilians-Universität in Munich and the Ifo Institute for Economic Research. All have been carefully selected and refereed by members of the CESifo network.

Introduction

Marc Gronwald and Beat Hintermann

Climate change is among the most important challenges we face today. Although the severity and distribution of its effects are not known with certainty, it seems clear that every part of the world will be affected in some way, and that the damages are likely to outweigh any potential benefits by a significant margin. Emissions trading schemes (also known as cap-and-trade markets) figure prominently among the policy instruments used to tackle this problem. The largest such market so far, the European Emissions Trading Scheme (EU ETS), started in 2005. Emission permit markets are also increasingly being used outside of Europe—for example, in China, in Canada, and in the United States.

The experience gained during the first years of the EU ETS, however, indicates that the task of establishing an emissions trading scheme in the real world, as opposed to in an economic model, is anything but simple. Evidence for the necessary learning process includes the allowance price crash observed at the end of Phase I, price volatility in general, and the apparent oversupply during the first two phases. Although the theory behind emissions markets is straightforward and states that they reduce emissions below a certain limit at the least cost, there are several complications that may arise in practice. Particular design features, the prevailing market structure, and interactions with other policy instruments can influence the efficiency of the scheme, in the sense that the abatement costs associated with attaining the emissions goal are not minimized. And because no obvious counterfactual exists, the reduction in emissions that can be attributed to the introduction of the EU ETS itself is not known, nor is its economic cost.

These issues are the subjects of a rapidly growing literature. A number of recent papers use empirical analyses of permit prices and their determinants to evaluate the initial trading periods of the EU ETS. The recent financial crisis motivated many researchers to focus on the role of macroeconomic factors as well as on the relationship between carbon markets and other financial markets. Other strands of the literature examine the optimal design of emissions trading schemes, the dynamic incentives of these schemes relative to those of alternative policy instruments, the interaction between concurrent policy instruments, the importance of market structure and allocation methods, and the opportunities and challenges associated with linking different permit markets.

To bring together theoretical and empirical economists who are actively contributing to research involving permit markets, we organized a workshop on "Emissions Trading as Climate Policy Instruments—Evaluation and Prospects" as part of the CESifo Venice Summer Institute (VSI). The goals of the workshop were to discuss recent findings, to help improve the design of future emissions trading schemes, and to identify relevant open questions for future research. We received a number of interesting contributions that analyzed some of the main aspects of permit markets. The workshop took place on July 22 and 23, 2013, and we hope that the participants share our view that it was a very interesting event that created a number of new insights. This volume contains the two keynote addresses of the workshop and eight of the nine contributed papers. (The last paper was presented at the workshop but was later withdrawn from the publication process.)

Topics covered

The eleven chapters are organized in four thematic parts. Part I describes the EU ETS from its start in 2005 until the present. Along with a brief description of the main features of the market (provided in chapter 1), it contains the workshop's two keynote lectures.

In chapter 2, which represents the first keynote lecture, Denny Ellerman gives an overview of the current state of knowledge on permit markets in general and on the EU ETS in particular. Ellerman argues that the permit markets that have been put in place so far appear to work in the sense that the caps were never exceeded and they produce a single, visible price for emissions reductions. However, it is much less clear to what extent these achieve their main objectives of reaching an emissions goal at least cost. Ellerman points out four things that are not yet well understood: the degree to which true opportunity costs are reflected in permit and output prices, the relationship between the allowance price level and the resulting quantity of emissions abatement, participants' banking behavior, and the microstructure of trading.

In chapter 3, which represents the second keynote lecture, Andreas Lange discusses the legal development of the EU ETS and the regulatory uncertainty arising from the rule changes that have been made over time. Since the start of the scheme, the method for allowance allocation, the sectors and countries encompassed by it, the limits on importing offsets, and other regulatory features have been changed, and more adjustments are likely to follow in the future. From a practical point of view, it is necessary to change the regulatory framework in response to new knowledge about abatement costs, the process of climate change, technological innovations, macroeconomic developments, or changing political conditions (e.g., the lack of a binding post-Kyoto global agreement). Because of commitment problems, declaring all relevant policy features to be fixed forever would not be credible. On the other hand, changing the policy environment leads to regulatory uncertainty, which has been shown to decrease the level of investment in renewable technology and to lead to other inefficiencies. Using a simple model, Lange shows that regulatory uncertainty can increase expected profits for emission-intensive firms, and that for these firms investing in renewable technology serves as insurance against high prices for emission permits.

Part II focuses on the political economy context of the EU ETS. In chapter 4, Oliver Sartor, Stephen Lecourt, and Clement Pallière examine allowance allocation in the EU ETS. Whereas during the first two phases allocation was mostly free and based on historic emissions, a significant share of the cap has been auctioned since the start of Phase III, and the remaining allowances are distributed on the basis of technological benchmarks. EU member countries had considerable leeway in deciding on the amount of free allocation they issued to firms in the first two phases, which raised the question of whether industrial interests were influencing allocation decisions. Sartor, Lecourt, and Pallière analyze the effect of the rule change for free allocation by industry and country and find a significant and positive relationship between firms' previous over-allocation and the reduction in their amount of free allocation for Phase III. This suggests that the new allocation rules eliminated a heterogeneity in free allocation beyond past emission intensities, which is an indication of successful lobbying efforts by firms.

In chapter 5, Philipp Hieronymi and David Schüller focus on the role of political economy in irreversible abatement decisions. In their framework, a portfolio of gas plants has to be gradually replaced, either with new gas plants or with wind turbines. In addition to this choice, firms can spend resources on lobbying to try to persuade the policy maker to relax the emissions constraint or to add a premium on wind energy. Lobbing is costly, and its outcome is uncertain. Hieronymi and Schueller study under which conditions firms lobby at all, how lobbying activity would affect investment decisions, and whether firms lobby for gas or wind turbines. Among the determinants are the level of the permit price, the effectiveness of lobbying, and the growth rate of permit prices. Their analysis suggests that considerations of political economy are important, especially in a system in which decisions about climate policy can be adjusted relatively easily. Another interpretation would be that, in order to prevent socially inefficient lobbying expenses and technology choices, climate policy should be designed so as to minimize the scope for lobbying efforts. Lobbying can therefore be viewed as another element in the tradeoff between flexibility and rigidity of policy.

Part III examines the implications of the coexistence of different instruments for climate policy. Economic theory states that the number of policy instruments should equal the number of identified market failures or (perhaps more generally) the number of different policy objectives. Although this rule is often violated in practice, the number of climate policy instruments seems especially large. Examples of policy instruments that overlap with the EU ETS include markets for renewable quotas, price floors in individual countries, taxes on energy use, feed-in tariffs for renewable electricity, and building standards.

In chapter 6, Peter Heindl, Peter Wood, and Frank Jotzo employ a model in which two countries provide a public good using a combination of price and quantity instruments. As a special case, Heindl, Wood, and Jotzo consider a cap-and-trade market that is supplemented with a tax. Focusing on the distribution of the costs of achieving a certain target, they show that, because of differences in marginal abatement costs, the cost of producing abatement increases in the country that introduced the additional tax and decreases in the other country, and that the overall costs of reducing emissions increase.

In chapter 7, Arnold Mulder differentiates between climate instruments that reduce the carbon intensity of firms covered by the ETS and climate instruments that are applied to non-ETS sectors (e.g., households) but reduce the demand for the products supplied by ETS firms.

Quantifying the aggregate effects of these instruments using a stochastic simulation model, Mulder finds that parallel instruments that reduce the demand for ETS products have a greater effect on the allowance price than instruments that reduce the energy intensity of ETS firms. Mulder estimates that the EU ETS becomes redundant if the combined effect of the parallel instruments is greater than 20–45 million tons of CO_2 per year.

The chapters in part IV are concerned with how firms actually responded to various features of the EU ETS. In chapter 8, Stefan Trück, Wolfgang Härdle, and Rafał Weron focus on the obvious yet sometimes overlooked fact that the EU ETS exhibits all the important attributes of a financial market: European Emission Allowances (EUAs) are traded on exchanges across Europe, banks as well as financial intermediaries play active roles, and there is a spot market as well as markets for derivatives such as futures and options. In this context, firms hedge against the carbon risk that is posed by fluctuating allowance prices. Trück, Härdle, and Weron use econometric techniques usually applied in the context of commodity pricing to analyze allowance prices and derivatives. They focus on the relationship between EUA spot and futures prices, which changed from "backwardation" (a situation in which futures prices are lower than the expected spot price) to "contango" (the opposite), and the presence of significant convenience yields, implying that carbon futures contracts are not priced according to the cost-of-carry model. They argue that the price behavior in the allowance market is considerably different from that in other commodity markets.

In chapter 9, Ralf Martin, Mirabelle Muûls, and Ulrich Wagner use data from interviews with representatives of 429 manufacturing firms in six European countries to investigate the extent to which firms act rationally in their abatement and trading decisions, as well as the importance of transaction costs. The interviews reveal that the majority of the firms did not trade in the market, but aimed to achieve compliance by means of their initial allocation of allowances. This finding contrasts with standard economic theory, which asserts that firms would sell any excess allowances on the market. It further implies that the pattern of free allocation affects the resulting abatement decisions.

The remaining two chapters focus on the link between the EU ETS and the process that was established by allowing firms covered by the EU ETS to cover some of their emissions using offsets from the Clean

Development Mechanism (CDM) and Joint Implementation (JI) provisions of the Kyoto Protocol.

In chapter 10, Nathan Braun, Timothy Fitzgerald, and Jason Pearcy develop a model in which firms can choose between two compliance assets, which are equivalent in terms of the right to emit but not in terms of the transaction costs associated with them. Also, the use of one of the two assets may be constrained, thus reflecting the institutional context of the EU ETS. The model of Braun, Fitzgerald, and Pearcy aims to explain the persistent price differential between EUAs and Kyoto offsets during Phase II and the considerable heterogeneity of offset use between firms. Comparing their model predictions with annual compliance data and rules about offset use from Phase II of the EU ETS, Braun, Fitzgerald, and Pearcy find that transaction costs explain a significant share of firms' choices with respect to abatement, permit trading, and offset use.

In chapter 11, Timothy Fitzgerald focuses on why firms did not use up their limit for offsets in the first four years of Phase II, insofar as they were cheaper than EUAs throughout the phase. He argues that since firms are limited in the number of offsets they can surrender, using (cheaper) offsets instead of EUAs is like exercising a limited spread option. Using offsets today may have a positive payoff, but using them the following year could be even more profitable, conditional on the development of the price spread. Firms are therefore confronted with the problem of when to exercise this option. Using a model of a compound European option and calibrating it using data from Phase II, Fitzgerald finds that the value of waiting (that is, the value of not exercising the option) exceeded the exercise value in all years of the phase other than the final year, which would explain the limited use of offsets during the first years.

Lessons learned and avenues for future research

Because this is a collection of individually written pieces that cover a range of subjects, no joint conclusion can be offered. Nevertheless, we believe that there are some lessons we can draw from the contributed chapters. We also propose some questions for future research. These questions are meant to complement the gaps in knowledge that Denny Ellerman identified in chapter 2.

First, the market works, at least in the narrow sense that emissions now have a price and the cap has not been exceeded. To what extent potential efficiency gains relative to non-market-based methods of emissions control have materialized is a different question, and one to which this book does not provide an answer. However, this is an interesting topic for future research. If the efficiency gains are found to be relatively small but the costs of setting up and operating an ETS are large, alternative policy instruments may be preferable. What seems clear is that firms were able to influence their national governments and to gain favorable treatment, as is evidenced by the pattern of free allocation during the first two phases. Besides leading to wealth transfers, rent seeking is known to lead to allocative inefficiencies. However, the EU responded to this problem by harmonizing allocation rules across member countries and by increasing the share of auctioning. In general, the EU ETS has been adjusted quite frequently to address problems that have been identified along the way. This is good news in the sense that the system can be adjusted; on the other hand, it introduces costs associated with regulatory uncertainty.

Second, governments should be careful about instituting additional policies to combat climate change. As is discussed in part II of this book, such policies can have a significant effect on the demand for permits, even if they are aimed at non-ETS sectors. If parallel instruments are introduced, their interaction with the EU ETS should be taken into account. For example, the introduction of feed-in tariffs and a renewable-energy quota can easily result in business-as-usual emissions falling below the cap, thus reducing the amount of necessary abatement and the permit price. To keep the incentives for market participants stable, the cap should be adjusted accordingly. The combined use of parallel instruments, even if its does not appear to be optimal from a theoretical point of view, is a reality that should be acknowledged by the academic community. Research investigating what policies could be combined to deliver the greatest benefit and how to administer them—a quest beyond the mitigation of greenhouse gases at least cost—could provide valuable policy recommendations.

Third, it seems that at least some market participants do not act as would be expected of fully rational, informed profit maximizers. Some firms chose not to sell surplus permits even when banking was not possible, and others did not access the market even though that would have been cheaper than achieving compliance on their own. Besides transaction costs, reasons for this "non-economic" behavior may include limited rationality, but also may include preferences that make firms and/or their customers view in-house abatement, purchasing permits from other ETS firms, and importing Kyoto offsets as qualitatively different, even though they all have the same effect on a firm's emissions constraint. On the other hand, many firms do appear to act according to standard theory, and at least some gains from trade in abatement efforts (which is what trading emission permits amounts to) seem to be realized. Again, the quantification of these gains are realized could be a question for future research, because they have to be compared against the administrative cost of running an ETS, and against alternative policies.

Some caveats should be noted. Because the chapters in this book analyze the EU ETS from an economic perspective, they are not interdisciplinary. But it is worth emphasizing that most of the authors do not hold to stringent assumptions when deriving theoretical properties of emissions trading, nor do their findings hinge on particular schools of economic thought or certain behavioral expectations. Rather, they analyze the EU ETS as it currently exists, and they use economic methods to do so. For example, chapter 3 acknowledges that the regulatory framework of the EU ETS is subject to ongoing political discussions and is thus not set in stone. Chapter 4 takes explicit account of the political context in which the EU ETS has been developed, and thus adopts a perspective that is far from the assumptions of neoclassical economic theory that often underlie economics papers. Likewise, chapter 7 focuses on overlapping instruments, which, from the perspective of neoclassical economics, should not exist unless other market failures are present. Chapters 8 and 9, and to some extent chapters 10 and 11, test some of the predictions implied by the assumption of fully informed and rational market participants, rather than taking these predictions as given.

Nevertheless, in view of the complex nature of the real world, assessing the effectiveness of a policy such as an emissions trading scheme is an interdisciplinary challenge. Wherever it is possible and fruitful, there should be an exchange between research from various fields. The EU ETS is no exception. In particular, the disciplines of law, sociology, political, and the natural sciences provide additional points of departure from which an emission permit system may be evaluated. Although it would be beyond the scope of this volume to extensively discuss contributions of other disciplines to emission permit markets, the following remarks briefly address a few issues that are under investigation in related disciplines and discuss links to economic analyses.

An important issue from a legal point of view is that carbon markets create new property rights, but their enforcement is affected by the tradeoff between stability (investor perspective) and flexibility (government perspective) discussed in chapter 3. This touches upon the fundamental EU principles of legal certainty and the protection of legitimate expectations required for planning by market players. (See Sharpston 1990 and Barrett 2001.)

Furthermore, recent sociological research has shown that the mass media provide a link between scientific research and the public. This "mediazation" of the sciences is attributable to the expectation of the public that researchers in universities and other institutions will report on their research activities (Rödder et al. 2012). Depending on the scientists' attitude to media visibility, this affects the choice of research topic and may also affect the nature of the published results. But the mass media also play an important role in how the political system observes the opinion of the general public (Luhmann 2000). Thus, there are complex interdependencies between sciences, the political system, and the media.

Political science also provides valuable insights—for example, into the formation of policy goals such as the emissions cap or participation rules. One relevant question is how information is transmitted from society to political decision makers. Klüver (2010) studies the role of special-interest groups in this context and finds that, because EU institutions are highly fragmented, information is transmitted relatively easily between the EU and interest groups. In addition, Kohler-Koch and Finke (2007) argue that a new governance strategy imposed in 2000 actively supports such information transfer. Information transfer is closely related to the considerations of political economy modeled in part II of this book. Furthermore, since climate change is a global issue, real progress will require the collaboration of independent nations—collaboration that might take the form of binding agreements or, in view of the failure to reach such an agreement for the post-Kyoto period at the Conference of Parties meeting in Copenhagen in 2009, might focus more on processes and cooperation among regions. In this context, input from political science is obviously crucial.

Economic analysis of emission permit markets abstracts from emissions damages in the sense that the cap is treated as given rather than set optimally on the basis of the expected damages of emissions. Deciding on the right cap requires input from the natural sciences. The reports published by the Intergovernmental Panel on Climate

Change summarize scientific evidence on carbon emissions and their implications for climate change. On the basis of damage assessments, which take the nonlinearities between emissions and damages inherent in feedback loops and threshold effects into account, it is difficult to reach the conclusion that the cap in the EU ETS is set adequately, because current allowance prices are an order of magnitude below the mean of the reported estimates for marginal damages and two orders of magnitude below the 95th percentile. (See, e.g., Tol 2005 or Downing et al. 2005.) At current allowance prices, few abatement-related investments are expected, and investments that have been made during periods of higher prices may turn out to generate a loss. It is therefore questionable whether the EU's goal of decarbonizing the economy is being achieved by means of the EU ETS. Studies by Ellerman and Buchner (2008) and Anderson and Di Maria (2011) suggest that some abatement probably occurred during the first two phases. Despite the allowance surplus that was measured ex post, both phases had periods of high allowances prices (see chapter 2), but the current low price level bodes ill for future investments in clean technology.

However, the narrow objective of a market for emission permits is to achieve a pre-defined emissions cap at least cost, and it may be that the EU ETS has delivered just that (or, more likely, it may have led to costs that are below those of realistic alternatives). Concluding that the EU ETS does not work on the basis of the discrepancy between marginal abatement costs and emission damages is like saying that a car malfunctions because it is driven at the wrong speed. After all, a low cost of reaching the cap is a feature, rather than a bug, of any ETS. Rather, we conclude that the market works (though not perfectly), and that, since abatement turned out to be cheaper than had been anticipated, we may want to purchase more of it by lowering the long-term cap of the ETS. Indeed, the European Commission is currently discussing increasing the rate at which the future cap is melted off, setting up allowance reserves, and introducing explicit price floors, any of which would tighten the relevant emissions constraint.

It is unrealistic to assume that policy makers will introduce measures that are truly optimal from the political, economic, societal, legal, and ecological points of view. Therefore, one should not demand too much of the EU ETS. In view of the desirable theoretical properties of emissions trading schemes relative to traditional approaches to emissions control, in view of the empirical evidence that emissions markets deliver at least some of the proposed efficiency gains, and in view of emissions markets' high degree of political acceptance relative to emissions taxes, we conclude that the EU ETS has been a success so far and that it will make a significant contribution to combatting climate change in a cost-effective way, especially if the emissions cap is tightened further and the system is expanded to include emissions from other sectors and other regions.

Editors' notes

Tons and tonnes
The chapters vary somewhat in the use of these units. We assume that a European who writes either 'ton' or 'tonne' means a metric ton.

EU membership
Because the exact meanings of the terms EU15, EU25, and EU27 may not be familiar to all readers, we list the countries that make up those groupings here.

EU15
Austria, Belgium, Denmark, Finland, France, Germany, Greece, Ireland, Italy, Luxembourg, Netherlands, Portugal, Spain, Sweden, United Kingdom

EU25
the EU15 (the old member countries) plus the ten Eastern European countries that joined the EU in 2004: Cyprus, Czech Republic, Estonia, Hungary, Latvia, Lithuania, Malta, Poland, Slovakia, and Slovenia

EU27
the EU25 plus Bulgaria and Romania

Acknowledgments

We thank Janina Ketterer, Sonja Peterson, Wilfried Rickels, Ulrich Wagner, and Luca Taschini for serving as discussants for the workshop and this book. We also thank CESifo for funding the workshop on San Servolo Island in Venice. It is difficult to think of a more pleasant place for hosting a scientific workshop. We further thank Olga Zudova and Rahel Aichele for valuable assistance with workshop logistics and the submission process of this volume, along with the anonymous referees and Emily Taber and John Covell of the MIT Press.

References

Anderson, B., and C. Di Maria. 2011. Abatement and allocation in the pilot phase of the EU ETS. *Environmental and Resource Economics* 48 (1): 83–103.

Barrett, G. 2001. Protecting legitimate expectations in European Community law and in domestic Irish law. *Yearbook of European Law* 20: 191–204.

Downing, T. E., D. Anthoff, R. Butterfield, M. Ceronsky, M. Grubb, J. Guo, C. Hepburn, et al. 2005. Scoping uncertainty in the social cost of carbon. Final Project Report, Social Cost of Carbon: A Closer Look at Uncertainty. U.K. Department for Environment Food and Rural Affairs and Stockholm Environment Institute.

Ellerman, A. D., and B. K. Buchner. 2008. Over-allocation or abatement? A preliminary analysis of the EU ETS based on the 2005–06 emissions data. *Environmental and Resource Economics* 41 (2): 267–287.

Klüver, H. 2010. Europeanization of lobbying activities: When national interest groups spill over to the European level. *Journal of European Integration* 32 (2): 175–191.

Kohler-Koch, B., and B. Finke. 2007. The institutional shaping of EU-society relations: A contribution to democracy via participation? *Journal of Civil Society* 3 (3): 205–221.

Luhmann, N. 2000. *The Reality of Mass Media*. Policy.

Rödder S., M. Franzen, and P. Weingart. 2012. *The Sciences' Media Connection—Public Communication and Its Repercussions*. Springer.

Sharpston, E. 1990. Legitimate expectations and economic reality. *European Law Review* 15: 105–107.

Tol, R. J. 2005. The marginal damage costs of carbon dioxide emissions: An assessment of the uncertainties. *Energy Policy* 33 (16): 2064–2074.

I   Current State and Development of the EU Emissions Trading System

1   The EU ETS

Beat Hintermann and Marc Gronwald

In this chapter we provide a brief overview of the European Union Emissions Trading Scheme, drawing from material in the EU's fact sheet (European Union Directorate-General for Climate Action 2013) and three directives (European Union 2003, 2004, 2009) and using data provided by the European Environment Agency and the EU Transactions Log (EUTL). For a more thorough review, which would be beyond the scope of this book, we refer interested readers to Ellerman et al. 2010 and Ellerman et al. 2015. The academic literature associated with this market is reviewed in detail in Hintermann et al. 2015 and in Martin et al. 2015.

1.1 Design and scope of the market

The EU ETS, the world's largest emissions permit market to date, covers about 45 percent of the European Union's CO_2 emissions. The currency of the market is the EU allowance (EUA), which provides the holder with a one-time right to emit one ton of CO_2 or other greenhouse gases with an equivalent heating potential. The institutional rules governing the market vary by market "phase." Phase I spanned the period 2005–2007 and was considered a test run for Phase II, which coincided with the Kyoto compliance period of 2008–2012. Pilot phase allowances could not be transferred to the second phase and lost their value if unused for compliance; however, since the start of Phase II, allowances can be banked for use in later phases. At present the system is in Phase III, covering the period 2013–2020.

The system applies to CO_2 emissions and equivalent amounts of nitrous oxide and perfluorocarbons from installations in energy-intensive industrial sectors.[1] The scope of the EU ETS has changed somewhat over time, as more countries have entered the system either by becoming EU members (Romania, Bulgaria, and Croatia) or by linking their national systems with the EU ETS (Norway, Liechtenstein, and Iceland; links with other systems are planned for the future). Because the threshold level of compulsory participation was reduced in some countries, the total number of covered installations has remained roughly constant at around 11,000. Since 2012, emissions from aviation have been included as well, although this sector has a separate emissions cap.[2]

Firms can trade allowances freely within the EU bilaterally, through brokers, or directly on a few commodity exchanges. In fact, anyone can open a "personal holding account" in a member country's registry and then buy and sell allowances; a significant share of allowance trades were handled by banks and financial institutions that themselves were not covered by the market but used EUAs as financial assets. By April 30 of each year, the registered firms have to surrender permits corresponding to their emissions in the previous calendar year. April 30 is also the date on which each installation's realized and externally verified emissions from the previous calendar year are made public by each EU country. There is a penalty for noncompliance for every ton of emitted CO_2 for which firms do not surrender an allowance. The penalty was €40 in Phase I, €100 in Phase II, and €250 in Phase III. In addition, these firms have to surrender the missing allowances in the following year. Although no borrowing is allowed between market phases, within a phase firms can borrow from the next year's free allocation, because they receive the allocation for the current year before they have to surrender allowances for the previous calendar year. Note that the no-borrowing constraint between phases is binding only if no allowances are banked; in view of the large allowance surplus (see below), firms can "quasi-borrow" from future phases by decreasing their planned amount of banked allowances.

Whereas allowances were distributed mostly at no cost during the first two phases, a significant share of allowances have been auctioned since the beginning of Phase III. Importantly, electricity producers in the EU15 no longer receive any free allocation at all. The share of auctioning is planned to be increased to 70 percent by 2020, and to 100 percent by 2027. The rules on free allocation have been changed for Phase III as well. Figure 1.1 shows the average level of free allocation and realized emissions during Phase II by country. The

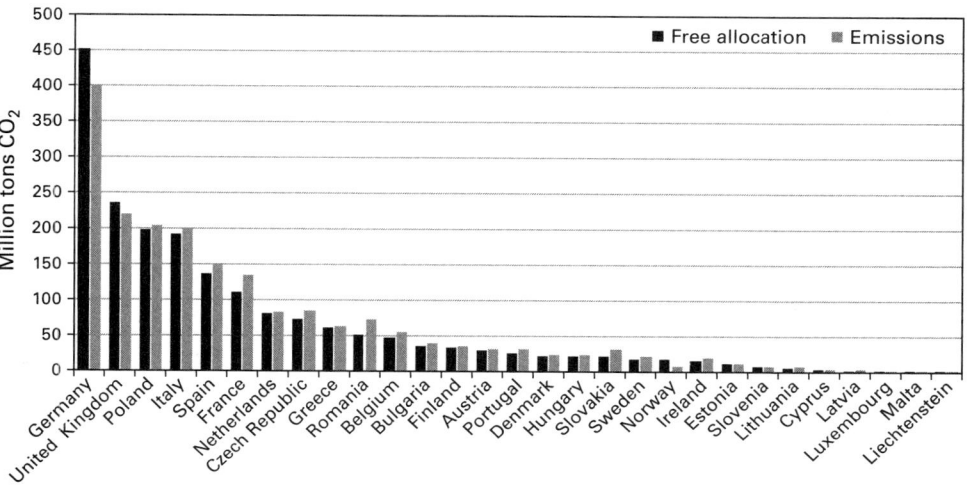

Figure 1.1
Annual allocation and emissions by country (average of Phase II).

system is dominated by Germany, the United Kingdom, Poland, Italy, Spain, and France, which together account for almost 70 percent of the EU's total emissions. Germany and the United Kingdom account for almost all of the allowance demand; most other countries are net sellers.

Free allocation and actual emissions by sector in Phase II are presented in figure 1.2. Combustion installations (i.e., plants that burn fossil fuels to generate power or heat) received nearly 67 percent of the total allocation and were responsible for 74 percent of aggregate emissions covered by the market. This was the only sector with a net shortage of allowances. All other sectors, on average, acted as net allowance suppliers.

Figure 1.3 provides an indication for the variance in the size of the included installations. About 69 percent of the covered installations are relatively small (less than 50,000 tons of CO_2 emissions per year), but together they emit only about 4 percent of total emissions. At the other end of the spectrum, the top 9 percent of the installations together account for 82 percent of emissions. Most of these large emitters are power plants. The largest annual emissions caused by a single installation were recorded by a power generator in Poland with 35.2 million tons in 2012, which exceeds the national average emissions of the 16 countries on the right side of figure 1.1.

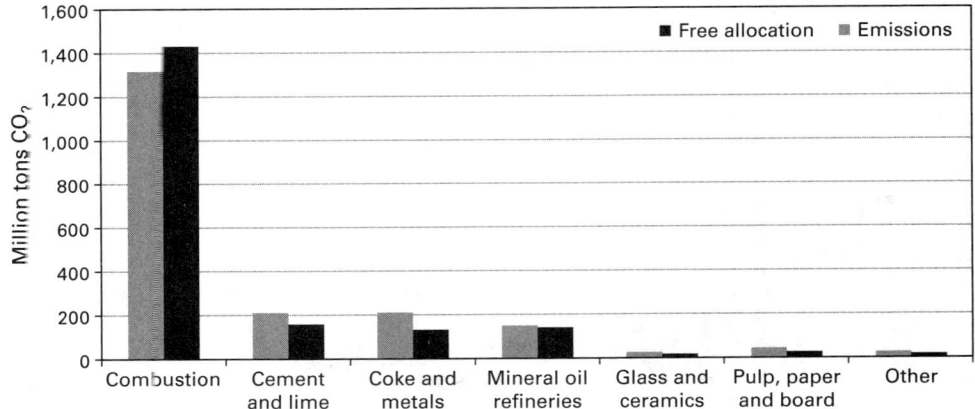

Figure 1.2
Annual allocation and emissions by sector (averages in Phase II).

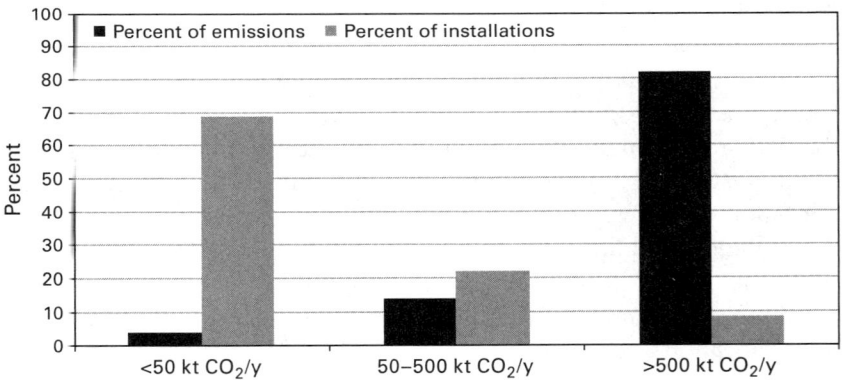

Figure 1.3
Distribution of allocation and emissions by size in Phase II.

1.2 Aggregate allocation and emissions

Figure 1.4 shows the aggregate emissions cap for the first three phases (free allocation plus auctions), along with verified emissions over the period 2005–2013; at the time of writing, emission figures for 2014 and later are not available. Starting in Phase III, the cap is "melted off" linearly at an annual rate of 1.74 percent relative to the average of the cap in Phase II. In addition to the cap specified in EUAs, firms were allowed to use a total of 1,418 million offset units from the Kyoto Pro-

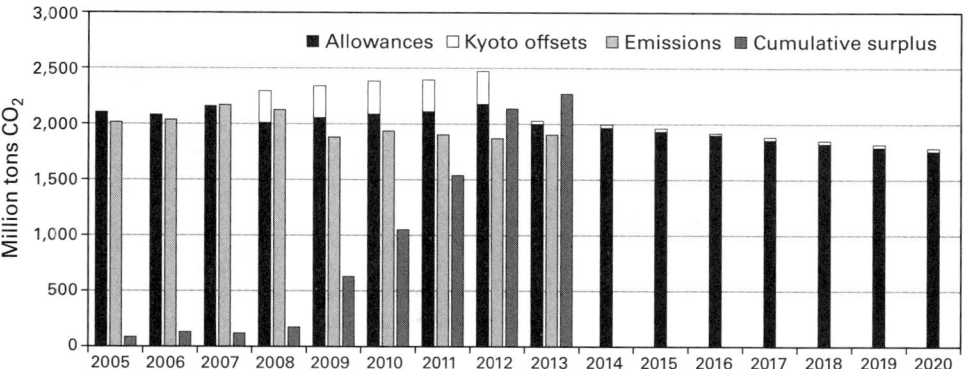

Figure 1.4
Allocation, emissions, and allowance surplus in the EU ETS, 2005–2013.

tocol flexible mechanisms during Phase II. Any unused import limits can be used in Phase III.[3] The use of these offsets by ETS firms is the subject of chapters 10 and 11 of this book. An additional import limit was set separately for Phase III, for firms that either received no free allocation or import allowance for Kyoto offsets during Phase II, or that significantly increased their capacity. The total amount of additional offsets that can be expected for Phase III is estimated to be around 160 million (Delbosc et al. 2011), bringing the total import limit to just below 1,600 million over the course of 13 years. Figure 1.4 also shows the cumulative allowance surplus through 2013. By 2013, the cumulative allowance surplus exceeded the total annual allocation.

There are several reasons for the significant allowance surplus. First and probably foremost is the financial crisis of 2008–09 and the subsequent banking and euro crises. Because of the fixed emissions cap, a reduction in business-as-usual (BAU) emissions lowers the need for emissions reductions by an equal amount. However, as shown in chapter 2, emissions per unit of output also decreased, implying that the recession alone cannot account for the decrease in emissions. BAU emissions also were reduced by the EU's additional policies aimed at increasing the share of renewable energy to 20 percent by 2020. For example, the feed-in tariffs implemented in Germany led to a significant increase in the generation of electricity by wind and solar power, a part of which came at the expense of fossil generation. (Generation by nuclear power also decreased during Phase II, _argely as a result of the political response to the accident at the Fukushima plant in Japan in March 2011.) The use of parallel instruments in the context of the EU's climate policy is discussed in chapters 6 and 7.

Another important reason for the surplus is the EU's decision to allow ETS firms to cover a limited portion of their emissions by using emission offsets generated by the Kyoto Protocol's Joint Implementation and Clean Development Mechanisms. These offsets have always been cheaper than EUAs, and their price declined to almost zero by the end of 2012; thus, it is optimal for a firm to use up its entire offset limit. This means that allowing firms to use 1,600 million Kyoto offsets in lieu of EUAs essentially increased the cap by this amount.

Another possible reason for the allowance surplus (perhaps a less obvious one) may be the way the emissions cap was set in the first two phases. The aggregate cap for Phases I and II emerged as the sum of individual caps in member countries' national allocation plans (NAPs). Since industrial emissions were not monitored before the start of the EU ETS, countries determined the Phase I NAPs by relying on emissions forecasts made in collaboration with the regulated firms. Because a higher forecast probably would result in more allowances' being allocated for free, firms had an incentive to report rather high emission forecasts and to lobby for larger country-level caps so as to gain an advantage over rivals located in other countries.[4] Firms' lobbying for a higher amount of free allocation is the subject of chapters 5 and 6.

To make matters worse, verified emissions recorded in 2005 were used to determine the cap and firms' free allowance allocation for Phase II. It is understandable from a practical point of view that the European Commission did not want to ignore information from the first actual emissions accounting in 2005. But because it had no other reliable baseline, making free allocation in Phase II conditional on actual emissions in Phase I conflicted with firms' incentives to abate emissions in the first phase. In effect, this allocation "updating" introduced a penalty for reducing emissions in the form of a reduction in future free allocation (Böhringer and Lange 2005; Harstad and Eskeland 2010). To the extent that the cap in Phase II was set proportional to verified 2005 emissions, an increase in the latter as a consequence of allocation updating would have led to an increase in the second-phase cap too.[5]

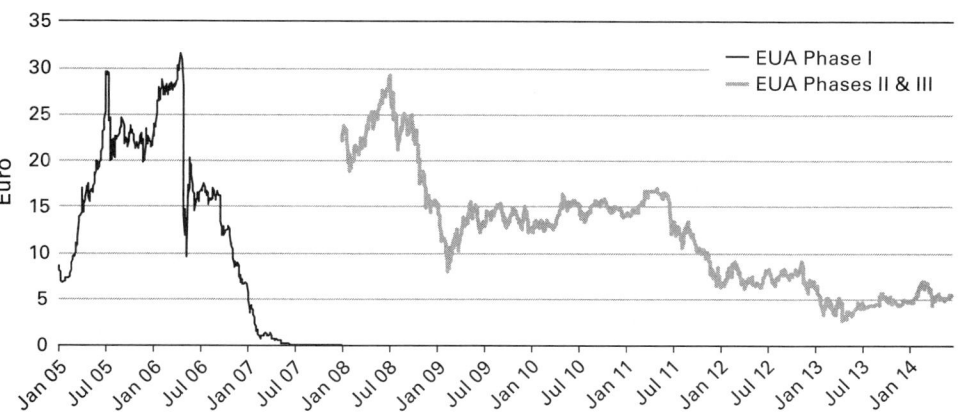

Figure 1.5
EUA price in Phases I–III. The price in Phase I is the over-the-counter price recorded by
Point Carbon. No single price is available for the price in Phases II and III that covers
the entire period; the price shown is a combination of a two-year future from Nord Pool
Spot (Europe's leading power market) until March 2008 and an end-of-year future from
European Energy Exchange (Germany's energy exchange) thereafter.

1.3 Allowance prices

Figure 1.5 shows allowance prices from January 2005 until June 2014.
Because of the no-banking provision, the EUA during Phase I was a
separate commodity from the EUA in later phases. The price crash in
April 2006 was a response to the first round of emissions verifications,
which revealed that emissions were well below the annual cap. A
growing literature has focused on price determination in the EU ETS;
for a review, see Hintermann et al. 2015. The fundamentals most often
associated with allowance prices are fuel prices, temperature, and economic activity. Because of the significant over-allocation, the price
decreased to zero at the end of the phase.

Because unused allowances in Phase II can be used to cover emissions in Phase III, the prices of these two market phases are linked by
arbitrage. The infinite time horizon introduced by unlimited banking
also decreased the importance of mean-reverting weather shocks,
because they are likely to be offset later by shocks of similar magnitude
but of opposite sign. The most obvious price fundamental is economic
activity. The financial crisis in 2008 led to a significant decrease in the
demand for allowances, which resulted in the sharp decline of the EUA

price from nearly €30 in spring 2008 to around €8 in early 2009. The price trajectories for the EUA during Phase I and since the start of Phase II imply that the cost of achieving the emissions target turned out to be cheaper than had been anticipated.

Notes

1. In Phases I and II, the participating sectors were labeled by ten activity codes, including the following: (1) combustion installations with a rated thermal input exceeding 20 mW; (2) mineral oil refineries; (3) coke ovens; (4) metal ore (including sulfide ore) roasting or sintering installations; (5) production of pig iron or steel (primary or secondary fusion) including continuous casting; (6) production of cement clinker in rotary kilns or lime in rotary kilns or in other furnaces; (7) manufacture of glass including glass fiber; (8) manufacture of ceramic products by firing, in particular roofing tiles, bricks, refractory bricks, tiles, stoneware or porcelain; (9) industrial plants for the production of (a) pulp from timber or other fibrous materials (b) paper and board; (10) installations from other industries that opted into the system. Since Phase III, a two-digit coding system has replaced this coding, and some additional sectors (e.g., manufacture of aluminum) have been included. More details can be found at http://ec.europa.eu/environment/ets/.

2. Allowances in the aviation sector are called EU aviation allowances (EUAAs). EUAs can be used by airlines in lieu of EUAAs, but not vice versa. The distinction between the "regular" cap and the aviation cap is presumably due to the desire to keep the former stable, whereas the latter will undergo changes once flights from and to non-EU countries are included in the system.

3. These offsets are certified emission reductions through the Clean Development Mechanism and emission reduction units from Joint Implementation within Annex B countries. Because of the "hot air" controversy, the EU did not allow the use of Assigned Amount Units.

4. With the overall cap held fixed, an increase in the national cap of one country at the expense of another country is equivalent to a transfer from the latter to the former, with the transfer value determined by the market price for allowances. Note that member countries' NAPs had to be approved by the European Commission. All but two NAPs were rejected in a first round and eventually adjusted downward, but not without a large number of court cases in which member countries tried (unsuccessfully) to defend their initial caps.

5. For a more thorough discussion of the merits and drawbacks of different forms of allowance allocation, see Hintermann and MacKenzie 2011.

References

Böhringer, C., and A. Lange. 2005. On the design of optimal grandfathering schemes for emission allowances. *European Economic Review* 49 (8): 2041–2055.

Delbosc, A., N. Stephan, V. Bellassen, A. Cormier, and B. Leguet. 2011. Assessment of supply-demand balance for Kyoto offsets (CERs and ERUs) up to 2020. CDC Climate Research working paper 2011-10.

Ellerman, A. D., F. J. Convery, and C. De Perthuis. 2010. *Pricing Carbon: The European Union Emissions Trading Scheme*. Cambridge University Press.

Ellerman, A. D., C. Marcantonini, and A. Zaklan. 2015. The EU ETS: Eight years and counting. *Review of Environmental Economics and Policy* (forthcoming).

European Union. 2003. Directive 2003/87/EC: Establishing a scheme for emission allowance trading within the Community and amending Council Directive 96/61/EC.

European Union. 2004. Directive 2004/101/EC: Amending Directive 2003/67/EC, establishing a scheme for greenhouse gas emission allowance trading within the Community, in respect of the Kyoto Protocol's project mechanisms.

European Union. 2009. Directive 2009/29/EC: Amending Directive 2003/67/EC so as to improve and extend the greenhouse gas emission allowance trading scheme of the Community.

European Union Directorate-General for Climate Action. 2013. The EU emissions trading system. Factsheet, EU Publications Office.

Harstad, B., and G. S. Eskeland. 2010. Trading for the future: Signaling in permit markets. *Journal of Public Economics* 94 (9): 749–760.

Hintermann, B., and I. A. MacKenzie. 2011. Reassessing the importance of initial allocation methods in emission permit markets. In *Advances in Environmental Research*, volume 12, ed. J. Daniels. Nova.

Hintermann, B., S. Peterson, and W. Rickels. 2015. Price behavior and market efficiency in Phase II of the EU ETS. *Review of Environmental Economics and Policy* (forthcoming).

Martin, R., M. Muûls, and U. J. Wagner. 2015. The impact of the EU ETS on regulated firms: What is the evidence after eight years? *Review of Environmental Economics and Policy* (forthcoming).

2 The EU ETS: What We Know and What We Don't Know

A. Denny Ellerman

In naming the workshop that led to this chapter Emissions Trading Systems as a Climate Policy Instrument, the organizers implicitly raised an important question: Are these systems appropriate instruments for climate policy? This question is timely not only because experience has been gained with the European Union's Emissions Trading System but also because of the marked de-emphasis of emissions trading as a climate instrument in the international negotiations under the United Nations Framework Convention for Climate Change (UNFCCC). These discussions have moved from the Kyoto Protocol (which many view as a failure), with its strong emphasis on emissions trading, to some new system of pledges and contributions with no special mention of emissions trading.

The EU ETS offers the most prominent example of trading in greenhouse-gas emissions. It is at once the world's first GHG ETS, its largest ETS for any type of emission, and the first multinational ETS. Accordingly, any discussion of the appropriateness of ETSs as a climate instrument must start with an evaluation of the EU ETS during its nine years of existence. My purpose here is simply to provide a perspective of what we know and what we don't know in attempting to make such an evaluation. I don't intend to enter into any of the arguments about the merits of an emissions trading system relative to a tax instrument or to conventional prescriptive regulation. Neither will I attempt a comprehensive evaluation of the EU ETS. What I wish to address are the empirical foundations for our knowledge about the EU ETS. In brief, my argument is that we know that the system works, but we don't know how or why it works. We have a theory to explain that, but empirical verification of that theoretical explanation is sadly lacking. Thus, what is developed here can also be seen as a generalized research agenda for the EU ETS.[1]

2.1 Emissions trading in the Kyoto Protocol and the EU ETS

Before proceeding, a few comments are appropriate concerning what is meant by emissions trading or emissions trading system as embodied in the Kyoto Protocol, the EU ETS, and the anticipated international agreement to be achieved at the UNFCCC's Conference of the Parties in Paris in 2015. Emissions trading in the Kyoto Protocol took a peculiar form unlike that typically labeled cap-and-trade. It looked like cap-and-trade in that certain signatories accepted quantitative limits on emissions and in that trading among signatories was authorized. It was more of what would now be called a baseline-and-credit system, but the most important distinction was that the agents trading were governments rather than firms as in conventional cap-and-trade programs, such as then existed in the United States' SO_2 Emissions Trading Program. Thus, trading under the Kyoto Protocol could not address any of the informational asymmetries that prevent non-market-based instruments from being efficient, or from achieving the desired emission limitation at the least cost. Signatories could have met their commitments by creating domestic cap-and-trade programs and linking them, but that was a matter for each signatory to decide. More generally, signatories could take whatever actions they wished to take in order to meet their commitments, including trading around the edges by the sale or purchase of assigned quota to the extent that domestic actions, or other events, might lead to over- or under-compliance with each signatory's Kyoto cap.

The EU ETS was inspired by the Kyoto Protocol, but its creators took the further step of creating a conventional cap-and-trade system in which the agents were firms, not governments. Member states had some leeway in setting their caps and in allocating allowances in the first two phases (2005–2007 and 2008–2012), but decisions about abatement actions were always made by firms, not governments. In these first eight years, the country caps were mandatorily downloaded almost entirely as free allocations to firms, which were then required to measure and report emissions and to surrender European Union Allowances (EUAs) equal to those emissions.[2] Until the start of the third phase in 2013, when an EU-wide cap was adopted and many EUAs were distributed by auction, the EU ETS could more usefully be seen as a set of coordinated but largely independent cap-and-trade systems with mandatory links among them, rather than as a single market.

What will be agreed under the United Nations Framework Convention on Climate Change in Paris in 2015 is yet to be determined as of the time of writing, but the broad differences from the Kyoto Protocol are clear. The agreement will be looser, broader, and more bottom-up than top-down. The signatories will agree to pledges or contributions to reducing global emissions more than commitments, and they will not be restricted to industrialized economies, as in the Kyoto Protocol, but will include at least the major emerging economies. The emphasis will be on what countries are willing and politically able to do, without any pretense of creating a global regime of legally binding commitments to which all would adhere sooner or later.

Emissions trading will not have the prominent place it enjoyed under the Kyoto Protocol, but it will not be absent—if for no other reason, owing to the presence of the EU ETS. Visions of a global emissions trading regime are still present, but it will emerge from a bottom-up construction through linkage among compatible systems, as national political circumstances and choices allow, instead of growing out of the top-down commitments as envisaged in the Kyoto Protocol. At least some of the initiatives anticipated in this successor agreement are meant to encourage cap-and-trade-like systems, perhaps at sectoral or sub-national levels, among which trading could be established. And just as trading systems are being established independently of the UNFCCC (for example, in California and Quebec), so the EU is pursuing linkages with other compatible trading systems independently, though carefully not excluding anything that might emerge from the 2015 agreement. In summary, emissions trading of the cap-and-trade form as embodied in the EU ETS remains very much alive as an important, if not the only policy instrument in an eventual global climate regime.

With this in mind, I now turn to what we know and don't know in attempting an evaluation of the EU ETS as a climate instrument. My focus will be on the EU ETS as a cap-and-trade system, as it might exist in any country, without any discussion of the important lessons to be drawn from the EU ETS concerning the creation of multi-national emissions trading systems.[3]

2.2 Issues that seem settled

Three things seem to me to be self-evident from the data and research on the EU ETS (as well as from other cap-and-trade systems): emissions are capped and reduced, markets form, and a single price emerges. These phenomena constitute the core of any cap-and-trade system. As such, they demonstrate that ETSs can achieve their basic objectives.

An ETS does limit and reduce emissions

Experience offers little support for claiming that ETSs fail to limit emissions to the cap. Covered emissions in the EU ETS were, cumulatively, 1.5 percent (80 million tons) below the cap in Phase I (2005–2007) and 6.2 percent (650 million tons) below in Phase II (2008–2012). However, the argument that is made against effectiveness of the EU ETS asserts that this outcome is a result of the economic recession and the ensuing euro crisis instead of the price for carbon. For sure, much of the reduction in the level of absolute emissions relative to pre-2008 levels is due to the recession; however, since then emissions have continued to decline while economic activity has experienced some recovery, as figure 2.1 shows.

The two top lines in figure 2.1 show the path of real EU25 GDP and industrial sector (including electricity) gross value added relative to 2004, the year before the start of the EU ETS. The bottom line shows the evolution of ET ETS emissions for installations that have been in the EU ETS from the beginning. The effect of the 2008 financial

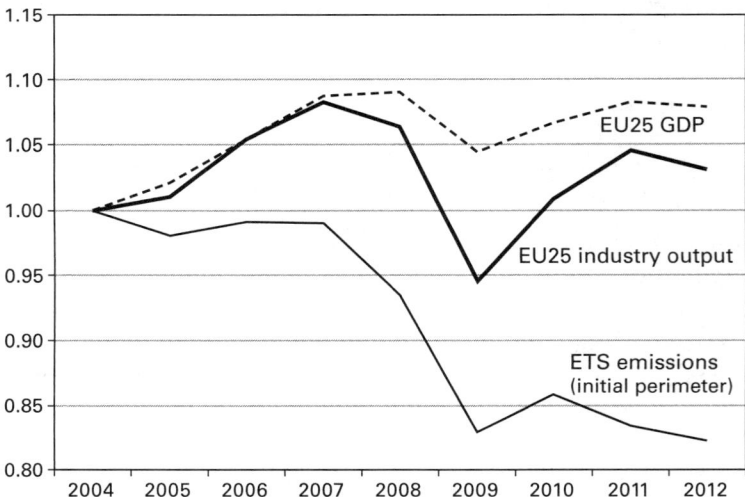

Figure 2.1
Indices of EU25 economic activity and EU ETS emissions (developed by the author on the basis of data from Eurostat and CITL).

crisis shows up clearly in all three indices. In the two years from the peak in 2007 to the trough in 2009, EU25 GDP falls by 4.0 percent, industrial gross value added (GVA) by 14.3 percent, and (same perimeter) EU ETS emissions by 16.2 percent. The real evidence of an independent effect of the ETS comes in the recovery. By 2012, GDP has recovered by 3.6 percent, industrial GVA is up 10.5 percent, but emissions are even lower (1.0 percent less than in 2009). Over the entire period from 2004 to 2012, real GDP increased by 7.9 percent and industrial GVA by 3.1 percent, but ETS emissions decreased by 18 percent. The 3.3 percent annual decline in emissions intensity relative to GDP is much greater than the 1 percent annual rate of decline observed before 2004, and more than can be accounted for by changes in industrial structure over those eight years or the effect of incentives for renewable energy (RE).

In the end, arguments about the magnitude of the reduction in emissions that can be attributed to the ETS depend on definitions of counterfactual emissions, something that cannot be directly observed. The value of the cap is that it limits emissions over the relevant time horizon to some predetermined level regardless of other factors that can cause counterfactual emissions to be higher or lower than initially estimated, such as rates of economic growth or fuel prices. When these exogenous factors differ from expectations, the effect is observed in the quantity of abatement performed to meet the always fixed cap and consequently the price. In the nine plus years of the EU ETS, lower-than-expected economic growth, probably the most important exogenous factor affecting emissions, has lowered counterfactual emissions, and thus reduced emission reduction requirements, consequently affecting the price of allowances.

Although the price of allowances is not zero, it is much lower than it was expected to be when Phase II began, and quite different from that seen at the end of Phase I, when the price did fall to zero (because of the inability to bank allowances from Phase I for use in Phase II and the irreversibility of abatement that had already been accomplished when the price of allowances was, for a while, as high as €30). Since banking is now allowed, the best evidence that the cap is binding (in the long run), has reduced emissions, and is expected to continue to do so is the evident unwillingness of holders of the nearly 2 billion banked allowances at the end of Phase II to sell these holdings at less than the price (€3–€4) that prevailed for part of 2013, and thus to drive the price to zero.

Allowance markets can be expected to form

It is hard to imagine now how dubious observers were that allowance markets would form when the first major cap-and-trade system, the SO_2 Emissions Trading Program, was set up in the United States in the early 1990s. It was a beautiful idea of economists, but one that would not work as a practical matter, especially for conservative, price-regulated electric utilities. The owners of affected facilities might trade within the confines of the firm to minimize cost, but inter-utility trading would occur only fitfully if at all. In fact, a market was slow to form. There were a few early trades, but a single highly visible (and lower than expected) price emerged as a result of early auctioning of a portion of the cap. The volume of allowances traded outside of the auctions was small at the beginning, but it grew quickly as the program got going and agents recognized that trading with other utilities with differing marginal costs would lead to further cost minimization.

When the EU ETS started, ten years later, there was no initial doubt whether a market would form, but the increase in the actual trading volume was similarly slow, as figure 2.2 shows.

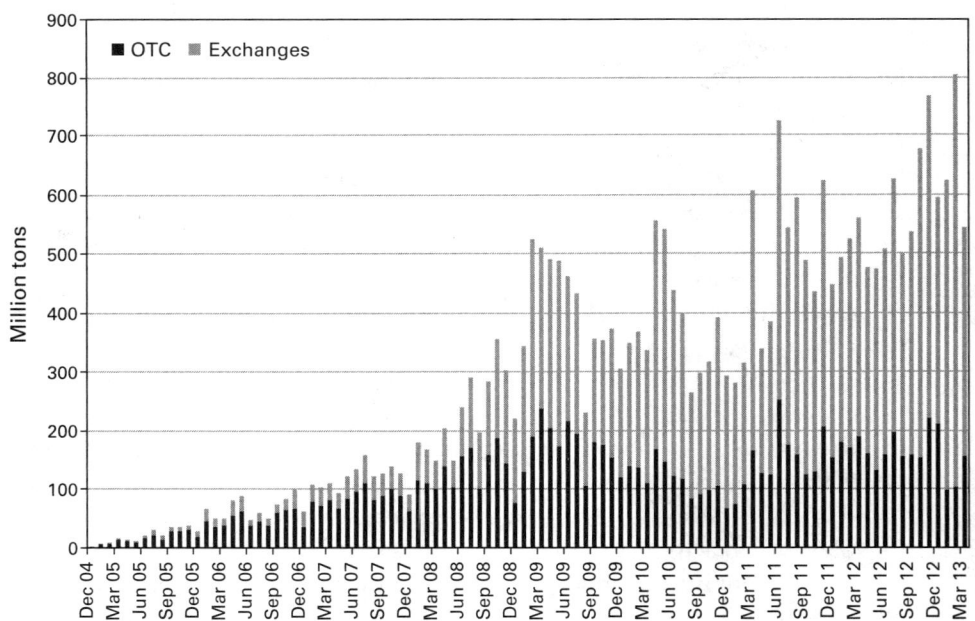

Figure 2.2
EUA monthly volume, 2005–2013 (data from Point Carbon).

Volumes of trading, initially small, grew steadily. At the end of Phase II, annual volumes are running at about 6 billion allowances, roughly three times the annual cap. In 2005, trading was almost exclusively over the counter, as it had been in the US SO_2 Emissions Trading Program and as it is in the Regional Greenhouse Gas Initiative in the northeastern US. At the beginning of Phase II, a more efficient trading mechanism—exchanges—began to take over, and EUAs traded much more like the equities, commodities, and other widely traded assets that these platforms handle. By the end of Phase II, five years later, exchanges were accounting for 80 percent of the observed trading.

Since a visible single price at any particular time is the prerequisite for equalizing marginal cost among the more than ten thousand installations included in the EU ETS, this condition for efficient and least-cost achievement of the limit on emissions is present. Whether all agents act on that price is another question, as will be discussed below; however, available data on the origin of surrendered allowances indicate that an EU-wide market has emerged. With only a few exceptions, allowances originating in any member state are surrendered in twenty or more other member states. This pattern suggests that intermediaries operate EU-wide instead of limiting themselves to trading in the particular member state in which they happen to be located.

A single price emerges

Perhaps no aspect of the EU ETS has received more attention, from both researchers and casual observers, than the price of EUAs. Prices were the first data available to researchers and the only aspect of the system that could be casually observed by others. The underlying question was (and remains) "Are the observed prices reasonable reflections of underlying realities determining demand, supply and the cost of abatement?" As figure 2.3 shows, the evolution of prices has given researchers and observers plenty to ponder.

Over the nine years of the EU ETS's existence, these front-December-futures prices have reached as high as €30, fallen to zero, then (in 2013, the first year of Phase III) settled in a range between €4 and €5. Before I describe the main movements, I should note that Phase I, the so-called trial period, was intentionally disconnected from Phase II. There was to be no banking from this initial phase to the one that followed (and no borrowing from the second phase, either). As a result, the final price would be either zero or the penalty price (€40) plus the price of the Phase II allowance, depending on whether final emissions were lower

Figure 2.3
EUA prices, 2005–2013 (weekly data from Point Carbon).

or higher than the cap. Initially, the prices for futures maturing in the two phases were similar because there was little basis to believe that Phase II would be more stringent than Phase I. However, once it became evident that a surplus in Phase I was likely, the two prices separated. Eventually the Phase I price fell to zero, while the post-2007 price rose as successive decisions on member states' caps indicated a more demanding Phase II cap than had been thought likely. Because banking was allowed and any allowances not used in Phase II could be used in later years, there was no separation of price between Phase II and Phase III as 2012 drew to a close.

The two main price adjustments—one in April 2006, one in late 2008—are well understood. The first, coinciding with the first releases of verified emission reports for 2005, was effectively the first possibility for calibrating expectations against real data. These first reports revealed emissions considerably lower than expected. The result was a sharp correction in price—by one-third for Phase II allowances and one-half for Phase I maturities in the space of one week. Thereafter expectations were calibrated; no subsequent releases of verified emissions data affected prices significantly. A similar marked decline in price occurred when the first emissions reports were released for the US SO$_2$ Emissions Trading, but over several months instead of a week.

The approximately 40 percent drop in allowance prices in late 2008 and early 2009 was due to the ongoing financial crisis, in the course of which most assets EUAs lost value. The subsequent further fall in EUA prices in 2011 corresponds to the period when the euro crisis was at its peak and it became evident that economic recovery in Europe would be later in arriving that had been hoped in the aftermath of the 2008 financial crisis.

The causes of the sudden initial rise in prices in 2005 were less obvious. Widespread initial expectations of EUA prices between €5 and €10 influenced the first trades, but prices rose quickly and stayed above €20 until the release of 2005 data allowed the initial calibration to take place. The price decrease in 2013 also remains unexplained.

Much has been written about pricing in the EU ETS. This literature is reviewed in a symposium article in the *Review of Environmental Economics and Policy* (Hintermann, Peterson, and Rickels 2015). While observing that the existing literature finds that the EUA price responds to various influences as would be expected, Hintermann et al. note that whether the price is fully efficient is not yet settled.

2.3 Issues yet to be analyzed seriously

Although the three phenomena discussed above constitute evidence that ETSs work, it cannot be assumed that agents and markets perform as flawlessly as economic theory would predict. The four topics discussed in this section—opportunity costs, the price-quantity relation for abatement, allowance banking, and the microstructure of trading—have received relatively little attention in the empirical literature, yet they are crucial to determining whether ETSs live up to their promise of being least-cost, and if not, by how much and why they fall short. These topics, concerning mostly how an ETS functions, constitute what we don't know about the EU ETS.

Opportunity cost

When an ETS distributes allowances by free allocation, the efficient, least-cost attribute of the ETS depends greatly upon the extent to which the recipients of allowances recognize opportunity cost. In the EU ETS, as in most cap-and-trade systems, allowances were distributed through free allocation based on historical conditions. Free allocation was reduced significantly with the beginning of Phase III in 2013, but an assessment of the efficiency of the system in the first two phases requires some assumption concerning whether agents recognized opportunity cost.

This issue is even more pertinent when a large number of firms receive a free allocation of allowances greater than actual emissions, as many industrial firms did in the EU ETS. These firms might well sell surplus allowances, but whether they have internalized the opportunity cost of the freely allocated allowances so that the price of carbon enters into production decisions is less clear.

Many firms argue that competitiveness effects, carbon leakage, and job losses are avoided with continuing free allocation, as if freely allocated allowances affect production decisions. Yet most of these firms are highly sophisticated, they clearly benefit *financially* from continuing free allocation, and they would benefit even more if they were to take opportunity costs into account in operational decisions. More nuanced views of the matter would argue that limits to management attention and rationality may cause opportunity costs to be overlooked. For instance, a major consideration in the proposal to substitute auctioning for free allocation in the EU ETS was a belief that free allocation dulls the sensitivity of agents to abatement opportunities. Others have suggested that, although firms may be alert to the value of excess allowances, internalizing the opportunity cost in search of internal abatement that would produce still more allowances is less frequently observed.

This is a difficult matter for researchers to address methodologically, as Fowlie and Perloff (2013) point out in their analysis of the NO_x RECLAIM ETS in California. If free allocation affects operational decisions, some relationship between allocation and emissions should exist independently of other factors. Data often suggest such a relation, but it can be shown to be spurious: allocations typically reflect historical emissions, and there is strong serial correlation between emissions from one year to the next. What is needed is some random or uncorrelated "break" in the data. In the EU ETS, such an opportunity may be provided by the complete cessation of free allocation to electric utilities in the EU15 beginning in 2013. If these agents do not recognize opportunity costs, there should be some discernible change in emissions in 2013 once all the other factors have been taken into account. A further complication is that the transaction costs of trading, especially the up-front fixed costs of trading, are considerably higher for small firms than for large ones. Thus, behavior that would appear to indicate a failure to recognize opportunity cost may be an indication for transaction costs, another form of inefficiency.

The issue is not a black-and-white one. It would not be surprising that some firms recognize opportunity costs and others do not. The important question is whether the latter constitute only a small minority of firms and emissions or whether such behavior is more widespread. If enough firms are failing to recognize opportunity costs, abatement will not take place where it is cheapest, and the price of allowances will be higher than it should be. But whether this is a serious distortion also depends on a second unknown.

The relationship between the price and quantity for abatement
Broad agreement exists, both in the public debate and the analytical community, that there is an inverse relationship between the price of allowances and the quantity of abatement. However, there is little agreement on the magnitudes of price or quantity effects of GHG abatement, and virtually no empirical research has been done on the subject. Every model can provide an estimate, but all these estimates are based on assumed elasticities in general equilibrium models or engineering-based estimates of the cost of various abatement technologies in linear programming models. Whatever agreement exists is based more on a common assumption than on any solid empirical data.

Yet many current policy issues depend on the magnitude of this relationship. For instance, the deployment of renewable generating capacity in the electricity sector certainly depresses the prices of EUAs, but whether that effect is great or small matters a great deal, especially when one is weighing it against other justifications for RE deployment. The current low EUA prices are attributed mostly to the effects of reduced economic activity, exacerbated perhaps by greater than expected RE deployment and offset use. Even if these latter influences are in line with earlier expectations, there is little doubt that EUA prices are lower as a result of these complementary policies, but no consensus on whether they are a little lower or a lot lower for this reason.

The dependence on modeling assumptions is not surprising, in view of the absence of a carbon price before the initiation of a cap-and-trade program and the consequent absence of data by which to estimate this relationship empirically. However, as Newell and Pizer (2008) show, this problem can be overcome, at least for abatement depending on the effect of the carbon price on absolute and relative fuel prices (in contrast to specific CO_2-abating retrofits). They examined fuel demand and equipment choices in the US commercial building sector to develop econometric results that they use to estimate the relationship for a carbon price of as much as US$100 per metric ton of carbon (not CO_2). Interestingly, when compared with the similar marginal abatement cost curves generated for the same sector by the Energy Information Agency's National Energy Modeling System (the main model used by the US government in estimating abatement costs), the effect of a carbon price on abatement was much greater. Their estimate indicated an 18 percent reduction in emissions in response to a $50 price instead of the 3 percent reduction indicated by the NEMS. Their work suggests considerably greater responsiveness in the relation of price and quantity of abatement than generally thought based on modeling assumptions. If applicable to the industrial and electricity generating sectors, this finding offers another possible explanation for the currently lower than expected prices in the EU ETS: model bias that underestimates the responsiveness of emissions to a carbon price.

Similar work focusing on the fuel demand effects of carbon pricing has been done for the EU ETS for the electricity sector using electrical system dispatch models. Delarue et al. (2010) show that these relationships are highly variable depending on the relative prices of coal and natural gas and the relationship between load and generation capacity. Generally speaking, abatement through fuel switching is limited by the amount of unutilized natural gas generation capacity that can displace online coal generation. Not only does this vary as load fluctuates by hour, day, and season; it is also influenced by fuel prices. Higher natural gas prices relative to coal create more unutilized natural gas capacity and therefore greater abatement potential through fuel switching. However, the same high relative natural gas price makes abatement more expensive so that a higher carbon price is needed to attain any given level of abatement. The converse is also true. Lower relative natural gas prices make abatement cheaper for any quantity, simultaneously reducing the abatement potential by putting more gas plants in service and thereby reducing switching possibilities.

Allowance banking
The greatest analytical failure with respect to the EU ETS is the failure to apply allowance banking theory. As a result, many in the policy world (and a surprising number in the academic world) attribute the existence of a large number of unused allowances at the end of Phase II to "over-allocation." The implications are that the cap was too generous, that emissions were going to be lower than the cap anyway, and

that the observed low price is an "option" on potential tightening of the cap.

The existence of banking cannot be gainsaid. The transition from Phase I to Phase II (in which banking was not allowed) demonstrated that prices reflect expectations. As it became evident that Phase I emissions would be lower than the three-year cap, the Phase I price went to zero in early 2007, and stayed there. In marked contrast, the Phase II price was always in the range €15–€20 even though the Phase II cap had not yet been defined. Clearly agents were looking ahead to the abatement requirements in Phase II and exchanging allowances within the range of expected Phase II prices.

After Phase II, when banking into Phase III was allowed, a large bank of nearly two billion allowances had been accumulated. A positive price prevailed, and there was no sharp break in price as Phase II transitioned to Phase III. Quite evidently, agents were unwilling to sell the "surplus" allowances held in accounts for a price less than the market price of €3–€5. This price was much lower than what had been expected at the beginning of Phase II, but expectations about future levels of economic activity had also changed. Although EU economic activity is now expected to be less robust than before, no one expects it to decline over the long term at a rate comparable to that of the ETS cap. Hence, expectations of future scarcity prevail as agents look ahead, as they had at the end of Phase I, and act accordingly. Although other explanations for the simultaneous existence of a large surplus and a positive price can be imagined, the simplest is banking.

This absence of attention to banking in the EU ETS is especially striking given the attention given to the subject in the US SO_2 Emissions Trading Program (a model for the EU ETS), in which emissions were also significantly lower than the cap in the first phase of the program. From the beginning, banking had been recognized as the explanation for the emission level below the cap—perhaps because there was a sharp discontinuity in the level of the cap after the first five years, in contrast with the smoothly declining cap of the EU ETS. Another explanation may be the immediate availability of registry data on allowance transfers instead of the five-year (now three-year) lag in the EU ETS. For the US program, Ellerman and Montero (2007) showed that the accumulation phase and first years of the draw-down of the banking period were what would be expected on the basis of independently derived discount rates and other features of that ETS when banking theory was applied.

Preliminary applications of banking theory to the EU ETS suggest that an end of Phase II bank of the magnitude observed is not unreasonable in view of low current interest rates and conservative estimates of future economic growth (Zaklan and Valero 2013). However, these same estimates indicate a banking period of more than thirty years, which seems longer than the effective economic horizon of agents. The effective economic horizon is not the only issue; there is also a question of the appropriate discount rate. In the US SO_2 Emissions Trading Program, the appropriate discount rate could be established as the risk-free rate. In the EU ETS, futures prices reveal a discount rate that seems independent of the euro risk-free rate.

There are other complications. For instance, is there a single discount rate when actors operate in different monetary systems, such as the euro, the pound sterling, and the Polish zloty? Moreover, the observed bank may not be entirely a result of active banking by agents; some portion of it may be due to passive banking by agents that, for one reason or another, do not sell allowances that they do not expect to need. Finally, there are unprecedented features of the EU ETS that might influence banking behavior, such as the sudden cessation of free allocation for electric utilities and the unexpected surge of very cheap offsets from Russia and Ukraine in the final months of Phase II. In the former case, greater exposure to the market and to the uncertainty of yet-to-be-developed auction procedures may have caused electric utilities to accumulate more allowances as a precautionary measure than they would have accumulated with continuing free allocation. In the latter case, the unexpected surge of very cheap offsets may have caused agents to end the period with more allowances than they had desired. And, in fact, prices moved down slightly in 2013, possibly due to the liquidation of unwanted end-of-year balances.

The microstructure of trading
Data on the transfers and holdings of EUAs by account holders have only recently become available, and preliminary analysis of these data (Zaklan 2013) reveal many things that heretofore could only be surmised. Chief among these is the relation between the trading of allowances on exchanges and the actual transfer of allowances among firms in the registries. Theory would suggest that trading observed on the financial markets is an indication for trading of allowances among firms, which is what minimizes cost and makes an ETS efficient. However, the quantities traded on the financial markets appear to be

much greater than those transferred among firms. The disproportion raises the question of whether the observed price is caused by trading motivated by differences between present and expected marginal costs, as is assumed in theory. This question arises because the spot and futures prices follow each other closely, because the greatest volume of trade occurs in futures, and because the principal participants in the buy side of the futures market are believed to be electric utilities that will readily assert that their purchases of futures are motivated not by speculation on future allowance prices but rather by the need to hedge price risk on forward electricity sales. This hedging behavior locks in the expected profit on the forward sale, which has been established by the present relationship among electricity, fuel, and allowance prices. Relatively little is known about the sell side of the allowance market. Speculators are surely present on both sides of the market, but the allowances that are committed to be delivered must eventually come from some firm with an expected abatement cost less than the offered futures price.

These and similar questions raise questions about the microstructure of EUA trading (that is, the identity and motivation of those participating in the markets where price is determined) and about how that behavior relates to the efficient transfer of allowances (the mechanism for equalizing marginal cost and attaining the efficiency that is an ETS's primary merit).

2.4 Issues that remain open

The four issues discussed in the preceding section have received little research attention. I would argue that little about them has been verified empirically. In this section, I briefly touch on three other issues that have received more policy and research attention, but which remain unsettled: offsets, effects on competitiveness, and market power. We can hope that continuing research will provide more definitive conclusions for the policy process.

The EU ETS is the first cap-and-trade program to make significant use of offsets. Most ETSs have offset provisions; however, because of the difficulty of determining baselines, monitoring emissions, and certifying the offsets, few if any offsets are used. The EU delegated these responsibilities to the procedures developed for the issuance of credits under the Kyoto Protocol's Clean Development and Joint Implementation mechanisms. The delegation was not open ended, as was evidenced by the quantitative limit placed on the use of these credits, the disallowance of certain types of projects, and the significant phasing out of offset use in Phase III. Still, the allowed use was large, and the high price of EUAs motivated project developers around the world—particularly in China, which is now implementing cap-and-trade pilot systems with an eye to linking with the EU ETS and other systems that may exist. At the same time, many questions have been raised about the additionality of some of these offsets and about their effect on the price of allowances in the EU ETS.

The effect of a carbon price on the global competitiveness of industries subject to an ETS is without doubt the most potent political argument used in opposing such systems and in proposing favorable treatment after their enactment. Surprisingly, what research has been done on the topic shows few effects, at least for the carbon prices experienced so far. The situation is similar to that in other environmental regulations for which the research on the pollution haven argument also fails to find strong supporting evidence. Yet the debate continues, and the issue cannot be considered to be settled. More research is needed both on the carbon price levels at which competitive effects occur and on the effectiveness of the palliative measures taken in the EU ETS (continuing free allocation) in mitigating these effects.

The potential for the exercise of market power in an allowance market has received much theoretical attention but little empirical analysis. In fact, market power is one of the feared consequences that have failed to appear in the EU ETS or in other cap-and-trade programs. The EU ETS certainly has been subject to tax evasion, phishing, hacking, and certified emission reduction units recycling, but none of these often-cited problems are problems of market power.

Conclusion

The EU ETS provides the best example of the use of a cap-and-trade system as a climate instrument. Whether such an instrument will play a large or a small role in the development of a global regime for addressing climate change will depend greatly on how the EU ETS performs. With nine years of experience, and without seeking to address political aspects of national and multinational systems, some conclusions can be drawn. My own view, based on close observation and an incomplete reading of the literature, is that the system works in the sense of limiting emissions and providing a price signal, but that we haven't gone very far beyond theoretical assumption in understanding how and why it works. For that reason, I would tilt the research agenda toward verifying these assumptions and coming to a better understanding of the role of opportunity costs, the price-quantity relationship for abatement, allowance banking, and the microstructure of trading. This is not meant to denigrate other topics, but these four constitute the topics that, in my view, deserve more attention.

Acknowledgments

The chapter on which this paper is based was commissioned by CESifo as a keynote address to the Workshop on Emissions Trading Systems as a Climate Policy Instrument at the 2013 Venice Summer Institute. I am indebted to Beat Hintermann and Marc Gronwald, the organizers of the workshop, for the invitation, and to them and other participants in the workshop for encouragement and for helpful comments.

Notes

1. Various assertions about findings are not referenced. References are made to articles of methodological interest on topics where research is needed. Those familiar with the field will readily recognize the sources and issues surrounding statements that I make. Also, readers interested in a comprehensive review of the economic literature on the EU ETS as it exists at the end of Phase II are invited to read the symposium of the EU ETS that is forthcoming in the *Review of Environmental Economics and Policy.*

2. Member states were given the option of auctioning up to 5 percent of the cap in Phase I and 10 percent of the cap in Phase II. In practice, relatively few member states chose to auction allowances, and only one (Denmark) auctioned the full allowed amount in the first phase. In the aggregate, less than 1 percent of the allowances distributed in the first phase were auctioned, and somewhat more than 3 percent in the second phase.

3. For a discussion of these issues, see Ellerman 2010.

References

Delarue, Erik D., A. Denny Ellerman, and William D. D'haeseleer. 2010. Robust MACCs? The topography of abatement by fuel switching in the European power sector. *Energy* 35 (3): 1465–1475.

Ellerman, A. Denny. 2010. The EU emission trading scheme: A prototype global system?" In *Post-Kyoto International Climate Policy: Implementing Architectures for Agreement*, ed. J. Aldy and R. Stavins. Cambridge University Press.

Ellerman, A. Denny, and Juan-Pablo Montero. 2007. The efficiency and robustness of allowance banking in the U.S. Acid Rain Program. *Energy Journal* 28 (4): 67–91.

Fowlie, Meredith, and Jeffrey M. Perloff. 2013. Distributing pollution rights in cap-and-trade programs: Are outcomes independent of allocation? *Review of Economics and Statistics* 95 (5): 1640–1652.

Hintermann, Beat, Sonja Peterson, and Wilfried Rickels. 2015. Price behavior and market efficiency in Phase II of the EU ETS. *Review of Environmental Economics and Policy* (forthcoming).

Newell, Richard G., and William A. Pizer. 2008. Carbon mitigation costs for the commercial building sector: Discrete-continuous choice analysis of multifuel energy demand. *Resource and Energy Economics* 30 (4): 527–539.

Zaklan, Aleksandar. 2013. Why do emitters trade carbon permits? Firm-level evidence from the European Emissions Trading Scheme. Working paper 2013/19, Robert Schuman Centre for Advanced Studies, European University Institute.

Zaklan, Aleksandar, and Vanessa Valero. 2013. EUA pricing: Banking or option value? Presentation at EUI Annual Climate Policy Research Conference.

3 EU Emissions Trading and Regulatory Uncertainty

Andreas Lange

The EU Emissions Trading System is a prominent example of how ideas on market-based environmental regulation have entered the political landscape and changed how policy makers and regulated firms view environmental policies. On the one hand, putting a price on carbon is seen as promoting investments into clean(er) technologies. On the other hand, the idea of reaching emissions targets at lowest costs by equalizing marginal abatement costs across sources has at last reached the policy stage after having been promoted by economists for several decades, beginning with Dales (1968), Montgomery (1972), and Tietenberg (1980).[1] Its final introduction in Europe in 2005 faced several important challenges. For example, decisions had to be made on the allocation of permits, on the choice and dynamic changes of the cap, on which sectors to include, and on how to deal with concerns of competitiveness and leakage. Putting such a large-scale emissions trading system into effect thereby also led to a new wave in the economics literature, including both theoretical and empirical studies (Convery 2009). How to adjust the emissions trading system in response to changing policy goals or to new economic conditions appears to remain a substantial challenge.

Changes to the EU ETS have been discussed prominently. The system was implemented in stages in order to also allow adjustments to the cap and to the allocation scheme as well as to allow sectoral extensions and possibly links with other international trading schemes (Böhringer and Lange 2013). Some of those potential changes had been envisioned from the beginning; others were triggered by observed market conditions. One example of the latter is a faster move toward auctioning in response to large windfall profits reaped by electricity providers in the first period (see, e.g., Sijm et al. 2006). Perhaps more important, the growing surplus of allowances as a consequence of over-allocation and the financial and economic crisis, among other things, caused the EU Commission to back-load emissions—that is, temporarily reduce the supply of allowances (EU Commission 2014a)—and to discuss the introduction of a market reserve to generate better price stability (EU Commission 2014b). Although these policy interventions and some changes to the rules of the EU ETS were triggered by the substantial breakdown of the allowance price, it is obvious that continued policy interventions and their interaction with the EU allowance market are crucial for the future price expectations and therefore may affect the incentives of firms to invest in cleaner technologies. This chapter discusses the policy dynamics that influenced the evolution of the EU ETS and how those dynamics may affect the efficiency of the regulatory system.

Several authors have studied dynamic interactions between market participants and regulators and indicated problems of time consistency in setting emissions caps (or also emission taxes). For example, Requate and Unold (2003) show how incentives that policy makers have to adjust their policies after investment in clean technologies has taken place affect the overall efficiency of emissions policies. Firms that anticipate such adjustments, however, may have distorted incentives to invest in the first place. Unold and Requate (2001) propose a system of call options with different strike prices to overcome the problem of asymmetric information between regulators and firms, and thereby to allow for temporal (endogenous) adjustments of the number of allowances in the market. While such studies have been motivated by potential misjudgments of carbon abatement costs at the time when the emissions trading scheme is implemented, less attention has been given in the literature to shocks to the economic conditions and how their anticipated effect on the regulatory environment affects investment decisions of firms.

Similarly, only a few authors attribute the choice of policy measures and later modifications to those measures to public-choice considerations and to governments' changing views of the importance of environmental issues. Notable exceptions are Gawel et al. (2013, 2014), who study the impact of political-economy considerations on the design of emissions trading schemes. Gawel et al. argue convincingly that the cap was influenced by interests of relevant sectors (see also Anger et al. 2008 and Markussen and Svendsen 2005) and that the necessary future adjustments to the cap leave room for continued political games and thereby create substantial regulatory uncertainty.

Hoffmann et al. (2009) define *regulatory uncertainty* as the perceived inability to predict the future state of the regulatory environment. Milliken (1987) further distinguishes among uncertainty about the future state of the environment of a firm, uncertainty about the future state of the firm itself (e.g., in the market environment relative to other competitors), and uncertainty about the consequences of decisions within this market environment. Reversing the causality between firms and regulators decisions, Fabrizio (2013) considers an index of regulatory uncertainty that is based on the "ability of political actors to alter the policy environment." Independent of the definition, uncertainty about future environmental policy may distort the incentives for firms to invest in new technologies.

In this chapter, I discuss the effect of regulatory uncertainty on the incentives provided by emissions trading systems. I first focus on the EU ETS as a changing regulatory environment. Next I summarize the potential effects of regulatory uncertainty on investment decisions. Finally, I use a simple analytical model to highlight the effect of uncertainty on firms' profits to gain insights into the relative effects on different technologies.

3.1 The EU ETS—a changing regulatory environment

The EU ETS was organized in phases. Phase I (2005–2007) was explicitly intended to be a trial period. As the EU ETS established the largest multinational system, both market participants and policy makers were expected to undergo learning. For that reason, changes to the system were already expected and even intended, since only with the beginning of Phase II (which was slated to begin in 2008) was the EU was committed to reaching its abatement target under the Kyoto Protocol. Some of the changes that were planned were related to the banking of allowances. The reason such banking was not permitted from the first phase into the second was that it was thought that it might generate too much supply in the second period, which might have made it difficult to achieve the Kyoto target. Banking was, however, envisioned to be allowed for future periods. Further adjustments related to the target itself. A stricter cap was chosen in the second trading period than in the first (about 6.5 percent below 2005 emission levels). Furthermore, the initial way of designing National Allocation Plans was already envisioned to be later replaced by more uniform allocation procedures, such as those relying on auctioning or benchmarking. However, as the exact rules were decided later, regulatory uncertainty was created. For example, the rules for benchmarking for the remaining free allocation were fixed in 2011, six years after the program began (EU Commission 2011).[2]

The design of the EU ETS changed substantially in Phase III, which—along with all planned future phases—was extended to eight years. National Allocation Plans were replaced by a centrally regulated allocation by the EU Commission. That change was motivated by a reduction target for the ETS segment of 21 percent below 2005 emission levels until 2020, which corresponds to the post-Kyoto milestone for EU climate policy established by the Climate and Energy Package. The cap was supposed to be decreased by 1.74 percent each year. It is envisioned that the cap will be decreased further after 2020.

From an economic perspective it may seem obvious that the EU ETS suffers from several flaws that limit its efficiency (Böhringer and Lange 2013):

• Because only part of the economy is covered, complementary measures have to be taken in non-ETS sectors. That may hinder achieving the overall abatement goal at least costs.
• Even at the beginning, it was obvious that a substantial oversupply of allowances would probably lead to a low emissions allowances price, whereas a sustained free allocation linked to production decisions may adversely affect the efficiency of abatement decisions by distorting production plans (Böhringer and Lange 2005a,b).
• The multitude of policy goals (e.g., renewable energy production) and corresponding policy instruments may also have adverse effects on efficiency.

As a consequence of these efficiency concerns, one could anticipate that further changes to the EU ETS will occur. The exact future specifications, however, remained uncertain and therefore could potentially also affect investment decisions.

Other changes were triggered by distributional concerns. For example, the recognition of large windfall profits to electricity providers in Phase I (Sijm et al. 2006) has contributed to the adjustment in the allocation method: whereas in Phase I all allowances were distributed freely, a move toward auctioning was implemented in Phase III, with full auctioning already beginning in the electricity sector.

The EU Commission promotes the ETS as "a major source of investment in environmentally sustainable development" and sees a sufficiently high carbon price as prerequisite in promoting investments in clean technologies (EU 2013). For this, it is necessary that firms react to the price in a rational way; hence, the price displays the current (and expected) scarcity of emissions allowances. Hintermann (2010) and Creti et al. (2012) show that the market prices were influenced by the market fundamentals, with some indication of undervaluing starting in 2009. The market price did, however, plunge to initially unforeseen levels. In Phase II, the price fluctuated from between €25 and €30 in 2008 to about €10 in 2012. Since then, the market has gone down again to levels around €5, sometimes even trading below €3. With this, the price is substantially lower than that of carbon traded in other cap-and-trade schemes, such as California's. The fluctuations in the allowance price can partly be accredited to a gradual recognition of over-allocation in the market and to increased use of offsets from the Clean Development Mechanism in 2011 and 2012 (Haita 2013), but changes in the overall economic environment and proposals to change the EU legislation clearly also contributed to the price adjustments. Haita identifies a substantial decline in the two days following the issuance of the Energy Efficiency Directive by the European Commission in June 2011, evidencing the importance of overlapping regulation aspects. The most important cause of the decline in the allowance price appears to be the economic downturn that followed the financial crisis of 2007. The reduced demand for allowances resulted in substantial overcapacity in the market—the EU Commission estimates a current surplus of almost 2 billion tons, approximately the allowances needed for a whole year.

As a reaction, the EU Commission proposed major interventions in the market. It immediately took steps to back-load emissions, i.e., to postpone the issuance of part of the allowances. In 2014, the auction volume is hence reduced by 400 million allowances (EU Commission 2014a). However, the direct effect of such back-loading should be minimal: as auctions are just delayed and not canceled and market prices should react to expectations in the available volume, the effect on price can be expected to be limited. For example, the back-loading decision led to reductions in the auctioned volume by 400 million allowances, as announced in February 2014. In line with the theoretically predicted minor effect, one could observe a temporal increase in the market price in February and March 2014 up to €7, while prices declined to January levels just below €5 by the end of March 2014. Naturally, this minor effect may have been driven by changes in other market fundamentals or may have been moderated by already incorporated expectations that such back-loading would occur.

Besides having a limited effect on the carbon prices, ad hoc modifications such as the back-loading decision may affect market participants' trust in the regulatory environment (Tol 2013). The backloading decision exemplifies how regulators may adjust the rules of the emissions trading system conditional on new information. While the attempt to support the carbon price given the goal of generating sufficient investment incentives into new technologies is plausible, alternatives to (potentially) stabilize the price have been discussed in the literature and have been implemented in existing cap-and-trade programs—for example, by officially announcing a price floor as well as a price cap, one could limit price fluctuations to a pre-specified corridor and thereby also generate a more reliable price signal. Philibert (2008) quantifies the benefits from implementing price ceilings and floors in the global climate architecture until 2050 and finds substantial decreases in expected costs and in the "breadth of economic uncertainty." The economic rationale behind setting price caps and floors lies in providing a more realistic and robust price signal, particularly through the presence of uncertain (marginal) abatement costs (Pizer 2002; Unold and Requate 2001). Price caps and floors may thereby alleviate the need to tighten or to loosen the initial emissions target in light of new developments.

The EU Commission also initiated a discussion of further structural reforms of the EU ETS. Recognizing that the large oversupply of allowances and the consequently low price reduces incentives for firms to invest, the EU Commission fears that the ability of the EU ETS to meet the ETS targets *within* future phases in a cost-effective manner will be limited (EU Commission 2014b). A sensible reform of the EU ETS would require that reliable price signals be generated in order to create incentives to invest in low-carbon technologies. Among the proposed measures are tightening the emissions cap (e.g., by decreasing the allocation until 2020 or by permanently retiring parts of the allowances), including additional sectors in order to strengthen demand, reducing the options to use offsets (CERs), and introducing a market reserve in combination with a price floor. Such a market reserve with a price floor would be a permanent measure—unlike the temporary back-loading—and is proposed to take effect in Phase IV, starting in 2021 (EU Commission 2014b). An administrative order was sufficient to allow back-loading, but establishing such a market reserve requires approval by all member states and by the European Parliament. Therefore, its prospects may be uncertain. The change would be accompanied by plans to increase the annual linear reduction factor from the current 1.74 percent to 2.2 percent starting in 2021 in order to achieve a 40 percent reduction target in the EU below 1990 emissions by 2030 in combination with complementary measures in sectors not covered by the ETS. Releasing 100 million allowances from the reserve (if they exist in the reserve) is proposed to be triggered when the number of allowances in circulation falls below 400 million. Without directly reducing price uncertainty, it limits the currently excessive gap between supply and demand.

Interestingly, the proposal to establish this market reserve in 2021 with the start of Phase IV was explicitly motivated by the attempt to "provide market participants with lead-time to adapt to the introduction of the reserve and sufficient regulatory certainty during phase 3 of the EU ETS" (EU Commission 2014b). The EU Commission explicitly refers to the goal of reducing regulatory uncertainty, while attempting to introduce changes to the system that may generate a more robust price signal by allowing the supply of EU allowances to fluctuate with changes in demand.

The introduction of a market reserve is a rather indirect way of affecting future allowance prices: it does not necessarily lead to a reliable price signal. If the goal is a robust minimum price, the already-mentioned concept of a price floor could be implemented by introducing a reserve price in the allowance auctions. As the current discussion on how to intervene in the allowance market was triggered by the low carbon prices, it would be beneficial to explicitly formulate a goal regarding the envisioned price range. Fell et al. (2012) demonstrate that most of the positive effects of market reserves by creating *soft* price controls are achieved when the reserve is modest, as is not likely to be the case under the current EU proposal in view of the current excess of allowances in the system. Perkis et al. (2013) demonstrate experimentally that hard price ceilings may be advantageous compared to soft ceilings as those implemented by a market reserve. In their laboratory experiment, they find that soft ceilings lack price control; they also find behavioral deviations from theoretical predictions that lead to distributional shifts in surplus from net buyers to net sellers. Their study is limited, however, to initially grandfathered permits. Furthermore, Stocking (2012) shows that the buffer stocks created by the reserve may also be subject to speculative attacks, thereby potentially

creating new incentives for market manipulations that limit the efficiency of the system.

   Because long-lived capital is involved in regulated sectors (particularly the energy sector), attempts to reduce the long-run uncertainty
would be laudable. However, formally committing to long-term
emissions-reduction goals and to indicators (e.g., prices, emissions,
technological advancements, economic conditions) that may lead to
adjustments to the policy may not be credible. Therefore, it appears
doubtful that the proposed changes to the EU ETS will create the
desired robust price signal and also a sufficiently reliable regulatory
environment.

3.2 The dynamics of firms' behavior under regulatory uncertainty

Not only does a functioning emissions trading system create price
signals to reduce emissions where it is cheapest; it also induces the right
incentives to invest in new technologies. Requate and Unold (2003)
compare several emission policy instruments and find that the investment incentives and the welfare ranking of instruments depend on the
option of regulators to commit to policy-levels and on their ability to
anticipate the investment strategies in new technologies. It is clear,
however, that a regime that does not allow for updating of the policy
in response to new developments in the economic environment will
fail to be efficient. As a reference case, Requate and Unold (2003)
assume that the regulator does not anticipate the new technology and
commits to an environmental policy based on the old investment
pattern. In that case, emission permits are found to lead to underinvest-
ment, and taxes usually create overinvestment. If the regulator can
commit to the policy level while anticipating investment patterns,
emission permits are found to outperform taxes and to create efficient
investment incentives. Among other cases, Requate and Unold also
study the situation in which the policy maker commits only to the
choice of the instrument (i.e., taxes or permit trading) but conditions
the choice of the tax rate or the cap on the observed investment decisions by firms. They find that both instruments may lead to socially
optimal investment levels. These results hinge, however, on several
important assumptions:

(i) The rules for specifying later targets are common knowledge:
observing the investment decisions and knowing the abatement costs

for all firms, the regulator minimizes the sum of abatement costs and social damages.

(ii) Firms can anticipate the future carbon prices and how they depend on the aggregate investment decisions.

(iii) Only two technologies with ex-ante known abatement costs are available.

(iv) Firms must invest in one technology or the other, and there is no possibility of delaying the investment decision.

In view of the described specifications and potential changes in the EU ETS, it is clear that assumptions i and ii are not satisfied, and that they are directly related to crucial features of regulatory uncertainty. Assumption iii is rather technical when combined with assumption iv, and the potential availability of new or improved technologies in the future may change the temporal investment patterns of firms. In what follows, I will concentrate on how regulatory uncertainty may affect the timing of investment decisions.

The notion of regulatory uncertainty has received substantial attention in the business literature. It is typically motivated by investments which are (at least partially) irreversible in that delaying investments may create a quasi-option value, as later decisions could be conditioned on new information (Pindyck 1991).

Fabrizio (2013) considers the effect of regulatory uncertainty on investments in renewable energy generation in the United States. She examines variations in state-level Renewable Portfolio Standards in the US electricity industry and shows that a state's history of policy reversals may trigger firms to reduce new investments. That is, regulatory instability may lead firms to fear that, because of future policy modifications, current investments may fail to pay off.

Guiso and Parigi (1999) also show that demand uncertainty weakens investments by slowing down capital accumulation for a sample of Italian manufacturing firms. Theoretically, Bloom et al. (2007) show that, because of partial irreversibility of decisions, uncertainty makes firms less responsive to demand shocks. Here, an increase in uncertainty raises the real option values and leads firms to invest/disinvest less. Teisberg (1993) demonstrates that firms may choose to invest in smaller projects with shorter lead times in the presence of uncertain environments.

Blyth et al. (2007) use a setup similar to that of Teisberg to explain how regulatory uncertainty, modeled as exogenous carbon price shocks, affects investments in coal-fired and gas-fired power plants and in carbon capture and storage (CCS). They show that uncertainty increases the risk premium for investments in power plants, though the risk premium is moderated by options of retrofitting with CCS (which serve as a hedge against changes in regulation levels).

It therefore becomes clear that firms do not necessarily postpone investment decisions, but rather adjust them. Typically, several investment options are available. As a consequence, more flexible projects with shorter lead times, or on a smaller scale, or with options to later adjust (retrofit) may offer comparative advantage. Looking for evidence from such adjustments, Hoffmann et al. (2009) survey 80 percent of the German power generation firms to investigate their responses to regulatory uncertainty within the EU ETS. They find that several firms stated reasons for *not* delaying investments: even in presence of regulatory uncertainty, firms may immediately invest in order to secure competitive or complementary resources, or to alleviate institutional pressure. In line with the work of Fabrizio (2013), Hoffmann et al. explicitly consider the option of firms to lobby in order to reduce the uncertainty. Engau and Hoffmann (2011) more specifically classify the available actions and distinguish among decisions to postpone investments, decisions to actively reduce uncertainties, decisions to adapt by restructuring the internal organization or the business portfolio, and decisions to disregard the uncertainties by following no-regret strategies or sticking to business-as-usual plans. They find that in the uncertain post-Kyoto world firms rarely pursue the strategy of postponing investments, but that regulatory uncertainty leads to major (welfare) costs as firms undertake costly adaptation measures or attempt to actively reduce the uncertainties through "contributing to the policy-making process." They call for more transparency in the political process and for reducing the level and the duration of major regulatory uncertainty.

Much of the literature on regulatory risk and investment decisions (e.g., Teisberg 1993) views the underlying uncertainty as asymmetric and focuses on the risk of expropriation. Considering investment options in energy generation under uncertain environmental regulation, however, induces risks as well as chances. Kipiyama and Suzuki (2004) use a real-options approach to show that nuclear power may profit from the introduction of (uncertain) carbon prices to electricity sector. Yang et al. (2008) compare coal-fired, gas-fired, and nuclear power plants and investigate how carbon price uncertainty and fuel price risk affect the relative performance of the respective technologies. They show that the risk premium crucially depends on which technology sets the price—i.e., on the merit order. On the one hand, carbon price risks are more important if coal-fired power plants sets the price, because of the large pass-through to the electricity price. On the other hand, nuclear power plants are more exposed to this risk, as for nuclear power plants it affects only the output price, whereas both costs and revenues are affected for the fossil-fuel-based technologies, which thereby reduces the effects of price variations on profitability. Similar arguments hold for renewable energy sources and other technologies whose emissions of greenhouse gases are less than those of conventional fossil-fuel based electricity generation.

The investments required for renewable electricity generation are often somewhat smaller than the investments required for fossil-fuel or nuclear power plants. Teisberg's (1993) argument that regulatory uncertainty may lead firms to invest in smaller projects may thus suggest some comparative advantages for investments in renewables.

Overall, one may conclude that increasing uncertainty usually distorts investment decisions, but that it affects different technologies to different extents. Whereas investing in some technologies (e.g., renewable energy) may be worthwhile only if the carbon price stays above some particular threshold value, the additional price uncertainty is clearly more relevant for emission-intensive energy sources on the cost side, while the revenue side (i.e., the cost pass-through) may alleviate this effect through changing revenues. For that reason, the effect of uncertainty on the comparative advantages of the different technologies when incorporating the timing of investments requires a detailed assessment that is beyond the scope of this chapter. In the next section, however, I will highlight the effect of uncertainty on the profitability of different stylized technologies without explicitly considering the initial investment decision.

3.3 The effect of carbon prices and uncertainty on profitability

To discuss the channels through which future carbon prices and regulatory uncertainty may affect the profit from different technologies, I set up a very simple stylized model inspired by the electricity market. I assume that different technologies can be characterized by the emissions intensity γ and the (uncertain) price of inputs $c(\gamma)$ to generate one kilowatt-hour of electricity. The merit order of the different technologies is additionally influenced by the carbon price, which for simplicity is assumed to be exogenous and is denoted by σ. That is, the total costs per kWh are given by

$$c(\gamma) + \sigma\gamma. \tag{1}$$

I assume that for each technology, investments have already taken place such that a capacity of $\kappa(\gamma)$ is available in the market. The demand Q may be uncertain, but can be satisfied through the total available capacity:

$$\sum_{\gamma} \kappa(\gamma) > Q.$$

Assuming perfect competition, the electricity price at a given moment in time is determined by the total unit costs of the marginal technology; i.e.,

$$\sum_{\gamma:p>c+\sigma\gamma} \kappa(\gamma) < Q \leq \sum_{\gamma:p\geq c+\sigma\gamma} \kappa(\gamma). \tag{2}$$

In other words, all technologies with total costs $c(\gamma) + \sigma\gamma$ less than the resulting electricity price are fully employed up to their capacity, while the marginal technology which sets the price delivers the remaining quantity.

This simple model allows to generate several insights: first, it is obvious that all technologies would benefit from lower own input costs $c(\gamma)$ and larger demand Q. The effect of an increase in the carbon price σ can be seen as follows: When it is assumed that there is no demand response to price changes, it follows from condition 2 that the cost pass-through is governed by the marginal technology $\hat{\gamma}$; i.e., $p = c(\hat{\gamma}) + \sigma\hat{\gamma}$ such that

$$\frac{dp}{d\sigma} = \hat{\gamma}. \tag{3}$$

Therefore, the profit level per kWh, $\pi = p - c(\gamma) - \sigma\gamma$, for any active firm changes according to

$$\frac{d\pi}{d\sigma} = \frac{dp - c(\gamma) - \sigma\gamma}{d\sigma} = \hat{\gamma} - \gamma \tag{4}$$

such that all active technologies with emissions intensity lower than the price-setting marginal technology will benefit from higher carbon prices. This applies to solar or wind power benefitting when prices are set by coal-fired power plants.

Though this insight is not surprising, it is important to put it into the perspective of the notion of regulatory uncertainty based on options to affect the regulatory environment. (See, e.g., Fabrizio 2013.) Whereas in Phase I of the EU ETS the excessive initial allocation and the low carbon price were partially attributed to extensive lobbying activities (Anger et al. 2008; Markussen and Svendsen 2005), the substantial presence of renewable capacity establishes political interest in keeping a robust and sufficiently high carbon price, which may counteract the interest in low carbon prices by emission-intensive industries.

The simple model sketched above also makes it possible to see the effects of price uncertainties. As usual, the effect of uncertainty on profits is defined as the difference when comparing the (expected) profit when the underlying variable (e.g., price or quantity) is random with the profit value that results when the random variable is replaced by its expected value. Naturally, the economic and regulatory environment may simultaneously affect demand Q and input prices $c(\gamma)$ as well as the carbon price σ. I will, however, consider these variables independently.

I start by considering how uncertain input prices affect technology γ. Here it is clear that for large $c(\gamma)$, the technology will not be active (given a specific level of demand and carbon price), for intermediate levels it may set the price such that again $d\pi(\gamma)/dc(\gamma) = 0$, whereas for low $c(\gamma)$ some other technology $\overline{\gamma}$ will be at the margin such that the costs changes do not affect the price such that $d\pi(\gamma)/dc(\gamma) = -1$. That is, the relationship between $c(\gamma)$ and $\pi(\gamma)$ is first linearly declining and then constant, i.e., convex. With Jensen's inequality, it therefore follows that uncertainty with respect to input prices has a positive effect on expected profits.

The effect of demand uncertainty is ambiguous. Again, for small Q, the technology may not be employed, while the curvature of the price development at higher demand levels will depend on the distribution of total costs and the installed capacity of other technologies. It may be (locally) concave or convex such that uncertainty with respect to the demand may decrease or increase expected profits.

In the case of the effect of carbon price uncertainty, it is instructive to first consider the effect on the output price p. It is clear that the marginal effect of an increase in σ on total costs per kWh for a given technology is γ. This implies, however, that for increasing carbon prices σ the marginal price-setting technology $\hat{\gamma}(\sigma)$ must become less and less emission intensive. This follows since $c(\gamma) + \sigma\gamma < c(\overline{\gamma}) + \sigma\overline{\gamma}$ for $\gamma < \overline{\gamma}$

implies that the same relationship holds when σ increases further. Therefore, the output price p is concave in the carbon price σ. As a consequence, the profit of an active technology—that is, a technology for which $\pi = p - c(\gamma) - \sigma\gamma > 0$—is also concave in σ, which indicates that uncertainty in the carbon price has a detrimental effect on the expected profits as long as this technology is active for all levels of the carbon price. In general, however, uncertainty in the carbon price may also benefit the profits from a technology if it is employed only for specific ranges of the carbon price. That is, for low (high) carbon-intensive technologies driven out of the market by low (high) carbon prices, the effect of uncertainty in the carbon price is ambiguous. The simple model thereby nicely illustrates the potential effect of uncertainty on renewable technologies relative to fossil-fuel-based electricity generation: a technology that would not be used at the average price (i.e., when no uncertainty exists) would benefit from uncertainty as this may lead to its (temporal) activation. Expected profits from technologies which are active for almost all possible prices, however, would be reduced due to the uncertainty. In particular, this may apply for "near-zero-emission" technologies such as solar or wind power as long as they operate at very small unit cost.

Coming back to the discussion of political interests implied by carbon policies, one may therefore predict heterogeneous interests across firms: those employing primarily renewable technologies, including solar and wind power, may be interested in *stable* and *high* carbon prices, while those using carbon-intensive technologies may very well be interested in fluctuating prices. However, it is clear that the time scale on which these prices fluctuate will matter for the perceptions of uncertainty by firms.

Though firms always face uncertainties in demand or in input and output prices, these uncertainties operate on small time scales. For that reason, most real-option approaches model such uncertainties based on Brownian motions with some specific time trend. Regulatory uncertainty, however, may more likely lead to structural breaks: unanticipated price jumps or discrete changes to the underlying price trends may occur at an unknown moment in time. One may therefore argue that political uncertainty, which is likely to result in less frequent but more substantial price changes, would impose qualitatively different risks to the profitability of firms. As major investments may turn out not to be valuable, necessary long-term investments may be affected. However, investments in low-carbon technologies may serve as insurance against large increases in the carbon price.

Conclusion

The nature of the climate problem requires that mitigation policies be pursued over a long time horizon. Over time, policy goals may change as new knowledge is generated on the sensitivity of the climate to accumulated emissions, but may also change as a result of new technological developments or changes in the global economic environment. Furthermore, the nature of political decisions, both at the national level and at the international level, creates substantial uncertainties with respect to future regulatory targets that individual firms may face. On the one hand, governments or the focus of policy may change. On the other hand, the structure of international cooperation in the post-Kyoto world is uncertain. Even in presence of these uncertainties, long-term decisions are required in many sectors of the economy in order to achieve the levels of mitigation that are seen as necessary to limit the adverse consequences of climate change. Investments in infrastructure or in power plants and decisions to develop new technologies serve as examples.

In this chapter I have discussed dynamic features of the current EU ETS that involve ad hoc changes in the number of allocated permits (back-loading) as well as major structural changes in reaction to the observed low carbon prices. When deciding about rule changes, policy makers should consider the tradeoff between the welfare gains from adjusting the policy to a changed environment and the losses that arise from regulatory uncertainty. Such uncertainty can affect the overall efficiency of reaching climate policy goals in different ways: they may reduce the incentives to invest, and they may change the temporal investment patterns in an undesired way. Although much of the literature concentrates on such negative effects, I argued that the relative effect of regulatory uncertainty on the available technologies is less clear. That is, if investments have to be made and delaying the decision is no option, regulatory uncertainty may actually increase the profits associated with particular technologies.

Though it is possible to identify reasons why short-term price uncertainty may negatively affect investments in renewable technologies, it also becomes clear that investments in such low-carbon technologies may serve as a safeguard against potential major increases in the carbon price. For that reason, the effect of regulatory uncertainty on investments remains ambiguous. Furthermore, in light of regulatory uncertainties, the option to operate some renewable technologies at scales smaller than that of fossil-fuel-based power plants may result in additional comparative advantages for renewable technologies.

Thus, I have argued that regulatory uncertainty is not, per se, detrimental to investment decisions, but rather that it has different effects on different technologies. It is clear, however, that regulatory uncertainty may result in inefficient investments, and that unnecessary uncertainty should be avoided.

It appears reasonable to try to commit to long-run policy goals and to specify indicators that form the basis for future policy adjustments—for example, to formulate policy goals conditional on economic development. This is particularly important because regulatory uncertainty poses the threat of long-lasting structural breaks that may go beyond the typical riskiness of dealing with fluctuating demand and fluctuating prices.

Notes

1. For an insightful discussion of the history of the idea of emissions trading in Europe, see Woerdman 2004.

2. See Commission Decision of 27 April 2011 determining transitional Union-wide rules for harmonized free allocation of emission allowances pursuant to Article 10a of Directive 2003/87/EC of the European Parliament and of the Council (notified under document C(2011) 2772).

References

Anger, N., C. Böhringer, and U. Oberndorfer. 2008. Public interest vs. interest groups: Allowance allocation in the EU Emissions Trading Scheme. Discussion paper 08-023, Zentrum für Europäische Wirtschaftsforschung, Mannheim.

Bloom, N., S. Bond, and J. van Reenen. 2007. Uncertainty and investment dynamics. *Review of Economic Studies* 74: 391–415.

Blyth W., R. Bradley, D. Brunn, C. Clarke, T. Wilson, and M. Yang. 2007. Investment risks under uncertain climate change policy. *Energy Policy* 35: 5766–5773.

Böhringer, C., and A. Lange. 2005a. On the design of optimal grandfathering schemes for emission allowances. *European Economic Review* 49: 2041–2055.

Böhringer, C., and A. Lange. 2005b. Economic implications of alternative allocation schemes for emission allowances. *Scandinavian Journal of Economics* 107: 563–581.

Böhringer, C., and A. Lange. 2013. European Union's Emissions Trading System. In *Encyclopedia of Energy, Natural Resource, and Environmental Economics*, volume 3, ed. J. Shogren. Elsevier.

Convery, F. 2009. Reflections—The emerging literature on emissions trading in Europe. *Review of Environmental Economics and Policy* 3 (1): 121–137.

Creti, A., P. A. Jouvet, and V. Mignon. 2012. Carbon price drivers: Phase I versus Phase II equilibrium? *Energy Economics* 34: 327–334.

Dales, J. H. 1968. *Pollution, Property and Prices: An Essay in Policy-Making and Economics.* University of Toronto Press.

Engau, C., and V. H. Hoffmann. 2011. Corporate response strategies to regulatory uncertainty: Evidence from uncertainty about post-Kyoto regulation. *Policy Sciences* 44 (1): 53–80.

EU Commission. 2011. 2011/278/EU: Commission Decision of 27 April 2011 determining transitional Union-wide rules for harmonised free allocation of emission allowances pursuant to Article 10a of Directive 2003/87/EC of the European Parliament and of the Council (http://eur-lex.europa.eu/legal-content/EN/TXT/PDF/?uri=CELEX:32011D0 278&from=EN).

EU Commission. 2014a. Commission Regulation (EU) No 176/2014 of 25 February 2014 amending Regulation (EU) No 1031/2010 in particular to determine the volumes of greenhouse gas emission allowances to be auctioned in 2013–2020 (http://eur-lex. europa.eu/legal-content/EN/TXT/PDF/?uri=CELEX:32014R0176&from=EN).

EU Commission. 2014b. Proposal for a Decision of the European Parliament and of the Council concerning the establishment and operation of a market stability reserve for the Union greenhouse gas emission trading scheme and amending Directive 2003/87/EC (http://ec.europa.eu/clima/policies/ets/reform/docs/com_2014_20_en.pdf).

EU. 2013. The EU Emissions Trading System (EU ETS) (http://ec.europa.eu/clima/ publications/docs/factsheet_ets_en.pdf).

Fabrizio, K. R. 2013. The effect of regulatory uncertainty on investment: Evidence from renewable energy generation. *Journal of Law Economics and Organization* 29 (4): 765–798.

Fell, H., D. Burtraw, R. D. Morgenstern, and K. L. Palmer. 2012. Soft and hard price collars in a cap-and-trade system: A comparative analysis. *Journal of Environmental Economics and Management* 64 (2): 183–198.

Gawel, E., S. Strunz, and P. Lehmann. 2013. Polit-ökonomische Grenzen des Emissionshandels und ihre Implikationen für die klima- und energiepolitische Instrumentenwahl. *Zeitschrift für Umweltpolitik und Umweltrecht* 36: 406–435.

Gawel, E., S. Strunz, and P. Lehmann. 2014. A public choice view on the climate and energy policy mix in the EU—How do the Emissions Trading Scheme and support for renewable energies interact? *Energy Policy* 64: 175–182.

Guiso, L., and G. Parigi. 1999. Investment and demand uncertainty. *Quarterly Journal of Economics* 114 (1): 185–227.

Haita, C. 2013. The State of the EU Carbon Market. Reflection 14/2013, International Centre for Climate Governance.

Hintermann, B. 2010. Allowance price drivers in the first phase of the EU ETS. *Journal of Environmental Economics and Management* 59: 43–56.

Hoffmann, V. H., T. Trautmann, and J. Hamprecht. 2009. Regulatory uncertainty—A reason to postpone investments? Not necessarily. *Journal of Management Studies* 46 (7): 1227–1253.

Kipiyama, E., and A. Suzuki. 2004. Use for real options in nuclear power plant valuation in the presence of uncertainty with CO_2 emission credit. *Journal of Nuclear Science and Technology* 41 (7): 756–764.

Markussen, P., and G. Tinggaard Svendsen. 2005. Industry lobbying and the political economy of GHG trade in the European Union. *Energy Policy* 33 (2): 245–255.

Milliken. F. J. 1987. Three types of perceived uncertainty about environment: State, effect, and response uncertainty. *Academy of Management Review* 12:133–143.

Montgomery, W. D. 1972. Markets in licenses and efficient pollution control programs. *Journal of Economic Theory* 5 (3): 395–418.

Pindyck. R. S. 1991. Irreversibility, uncertainty, and investment. *Journal of Economic Literature* 29 (3): 1110–1148.

Philibert, C. 2008. Price caps and price floors in climate policy—A quantitative assessment. IEA Information Paper.

Perkis, D. F., T. N. Cason, and W. E. Tyner. 2013. An experimental investigation of hard and soft price ceilings in emissions permit markets. Working paper, Purdue University.

Pizer, W. A. 2002. Combining price and quantity control to mitigate global climate change. *Journal of Public Economics* 85: 409–434.

Requate, T., and W. Unold. 2003. Environmental policy incentives to adopt advanced abatement technology: Will the true ranking please stand up? *European Economic Review* 47: 125–146.

Sijm, J., K. Neuhoff, and Y. Chen. 2006. CO_2 cost pass-through and windfall profits in the power sector. *Climate Policy* 6 (1): 49–72.

Stocking, A. 2012. Unintended consequences of price controls: An application to allowance markets. *Journal of Environmental Economics and Management* 63 (1): 120–136.

Teisberg, E. Olmstead. 1993. Capital investment strategies under uncertain regulation. *Rand Journal of Economics* 24 (4): 591–604.

Tietenberg, T. 1980. Transferable discharge permits and the control of stationary source air pollution: A survey and synthesis. *Land Economics* 56 (4): 391–416.

Tol, R. 2013. Carbon tax: Still the best way forward for climate policy. *Inter Economics* 48 (2): 70–71.

Unold. W., and T. Requate. 2001. Pollution control by options trading. *Economics Letters* 73: 353–358.

Woerdman, E. 2004. Path-dependent climate policy: The history and future of emissions trading in Europe. *European Environment* 14: 261–275.

Yang, M., W. Blyth, R. Bradley, D. Bunn, C. Clarke, and T. Wilson. 2008. Evaluating the power investment options with uncertainty in climate policy. *Energy Economics* 30: 1933–1950.

II The Political Economy of the EU Emissions Trading System

4 Benchmark-Based Free Allocations in EU ETS Phase III: How Much Better Than Phase II?

Oliver Sartor, Stephen Lecourt, and Clement Pallière

One of the most controversial aspects of the European Union Emissions Trading Scheme during its first eight years has been the question of whether or not, and if so then how, to initially allocate emissions allowances to covered installations. Despite the early theoretical result of Montgomery (1972), who argued that initial allocation decision should matter for welfare but not for the economic efficiency of an ETS, a substantial literature has emerged since the creation of the EU ETS debating the relative merits of alternative initial allocation approaches. This chapter contributes to the literature by presenting a detailed empirical analysis and evaluation of some of the welfare and efficiency properties of the new free allocation rules for non-electricity sectors in the EU ETS in Phase III (2013–2020). It exploits an original database compiled by the authors to focus on the question of harmonization of allocations across the EU, which was one of the main points of the criticism of allocation methods in Phase I (2005–2007) and Phase II (2008–2012). It asks whether or not the introduction of harmonized benchmarks adequately addresses the potential for distortions to both the carbon market and internal product markets that was discussed in the earlier literature. It is shown, via the example of the cement sector, that considerable scope remains for internal market distortions despite certain improvements in Phase III. On the other hand, benchmark-based allocations are found to be a significant improvement on Phase II National Allocation Plans (NAPs) in terms of their distributional welfare consequences in several ways.

The literature on the initial allocation of pollution licenses has generally viewed free allocation as (at best) a second-best approach to initial allocation in the presence of risk of carbon leakage. This view has been based both on arguments about distributional welfare and on arguments about economic efficiency. The idea that initial allocations can have important distributional implications goes back to Montgomery (1972), who demonstrated that free allocations have welfare consequences because of the distributional effects of the initial license endowments, but that in the absence of transaction costs free allocations should not change the efficient market outcome. Since Montgomery, our understanding of other kinds of welfare implications of grandfathering allowances has improved. Bovenberg and Goulder (2002), Sijm et al. (2006), and Smale et al. (2006) have shown that, where companies pass on the opportunity costs of pollution licenses into consumer prices, 100 percent free allocation leads to windfall profits for polluting industry. Furthermore, Crampton and Kerr (2002) and Smale et al. (2006), Åhman et al. (2007), Palmer et al. (2006), Kruger et al. (2007), and Benz et al. (2009) have argued that a higher share of initial auctioning is better for aggregate welfare, all else equal, since it avoids the emergence of the initial division of allowances.

In the context of current climate-change policy, in which carbon prices are unequal in different parts of the world for sectors competing in global markets, free allocation to carbon-intensive sectors is sometimes considered a second-best option for mitigating carbon leakage (Smale et al. 2006; Ellerman et al. 2007, 2010). Against this, a number of authors have argued persuasively that a combination of full auctioning and a border carbon adjustment (BCA) is, in principle, the most efficient and environmentally effective way to mitigate leakage (Monjon and Quirion 2010). But since BCAs are often thought to be diplomatically difficult to implement, much of the literature still focuses on how to make free allocation more equitable and more efficient.

The much- criticized decentralized National Allocation Plan approach taken in Phases I and II has illustrated the importance of ensuring that such "second-best" approaches to free allocation be applied well. The criticisms of the NAPs partly reflected distributional welfare concerns about the sheer levels of free allocation given to industry. For example, Trotignon and Delbosc (2008) found evidence of significant over-allocations of allowances to non-combustion sectors of the EU ETS in Phase I, ranging from 104.2 percent to 120.3 percent of actual emissions. Pearson (2010) calculated that even before the economic crisis of 2009 similar levels of over-allocations in most important sectors continued in Phase II, with the potential for windfall profits in certain sectors. Studies examining the degree of abatement in the non-electricity sectors have also shown that this level of over-allocation can almost entirely be attributed to high allocations, rather than significant abatement (Abrell et al. 2011).

A growing body of literature also suggests that free allocation can have real effects on economic efficiency of the emissions market outcome. Abrell et al. (2011) have shown, using firm-level data on EU ETS installations, that both in 2005 and in 2008 firms that were allocated a lower ratio of free allowances relative to their emissions tended to reduced their emissions in the subsequent year much more than those with higher free allocation rates. On average, firms that were allocated less than 80 percent of their 2007 emissions in 2008 reduced their emissions, on average, by 6.3 percent, whereas firms allocated more than 80 percent of their 2007 emissions in the same year did not reduce their emissions at all. Similarly, in Phase I firms that were allocated less than 115 percent of emissions in the Phase I NAP reduced their emissions by 3.4 percent on average in 2006, whereas firms allocated more than 115 percent of their emissions achieved no statistically significant reduction. Although the thresholds used in the study are somewhat arbitrary, the results suggest that, contrary to opportunity cost theory, the level of allocation can have a considerable effect on abatement levels (perhaps by affecting the level of management attention dedicated to abatement options) and can therefore be a source of inefficiency.

Several authors have focused on the possibility for distortions in final product markets in Phases I and II. Specifically, free-allocation methods under the NAPs were also found to be poorly harmonized across member states. As member states exercised considerable discretion about allocations to installations on their own territories, the risks of intra-EU competitiveness or investment location distortions were found to be high (Betz et al. 2004, 2006; Betz and Sato 2006; del Rio Gonzalez 2006). These effects were made worse by emissions baseline inflation, as identified empirically by Neuhoff et al. (2006) and Anderson and Di Maria (2011). Even within member states, the combination of grandfathering and windfall profits in the power sector was found empirically to be distortionary for investments in new power plant capacity by Pahle et al. (2011) and was shown to be possible in a theoretical model by Golombek et al. (2013). Thus, although carbon leakage risks seemed to have been effectively mitigated—Ellerman et al. (2010) and Sartor (2012) found no statistically significant evidence of leakage occurring during EU ETS Phases I and II—potential for intra-EU competitiveness and abatement-cost distortions remained.

Several studies have therefore begun to look at how best to initially allocate allowances to mitigate carbon leakage risk. Demailly and Quirion (2006), Böhringer et al. (2010), and Golombek et al. (2013) all showed that output-based allocations better mitigate carbon leakage and leave less scope for windfall prof its. However, they do so at the cost of reducing economic efficiency and potentially the environmental effectiveness of the carbon market by eliminating the channel of low-carbon product substitution. Martin et al. (2012) note that, if the goal of free allocation is to mitigate carbon leakage, then the optimal free allocation level should equate the marginal costs of issuing free allowance to the marginal benefits of mitigating carbon leakage at each installation and they present evidence suggesting that the Phase III EU ETS allocation rules do not do so as well as they could. Clò (2010) and Dröge and Cooper (2010) analyze the trade-exposure criterion for determining if a sector is at risk of carbon leakage and determine that it unnecessarily inflates the list of sectors considered at risk. This chapter contributes to the latter literature, which seeks to evaluate ways in which free allocations, if they are to be used in the context of the EU ETS, might be improved. It seeks to contribute to the existing literature on optimal free allocation policy by showing empirically how an ex-ante benchmarking system looks in practice from the standpoints of welfare and efficiency. It focuses on the extent to which Phase II allocations were potentially distortionary to the internal market for energy-intensive products in the cement sector and on the extent to which the introduction of harmonized free allocations actually do and don't reduce these distortion risks. In addition, it presents a series of stylized facts that serve to highlight some of the important ways in which Phase III allocation rules have concretely improved the welfare and efficiency properties of free allocations in the EU ETS.

4.1 The new benchmarking rules

In the first two phases of the EU ETS, over 90 percent of allowances were allocated free of charge to installations by means of a decentralized system whereby each member state drew up its own national allocation plan and emissions were roughly allocated according to historical emissions or capacity (Ellerman et al. 2010). The basic formula that determines each installation's allocation for each of its eligible products can be summarized as follows (EC 2011)[1]:

$$FA_{i,p,t} = BM_p \times HAL_{i,p} \times CLEF_{p,t} \times CSCF_t, \tag{1}$$

where $FA_{i,p,t}$ is the total free allocation that installation i receives for its product p in year t. BM_p, the product emissions-intensity benchmark of product p, is typically measured in tons of CO_2 equivalent per unit of output, and is typically based on the average emissions intensity of the most efficient 10 percent of installations in the EU ETS in 2007–08, although fallback approaches exist when this is not technically possible to evaluate or the measure is considered inappropriate as a sectoral benchmark.[2] $HAL_{i,p}$ is the reference historical activity (production) level of product p by installation i, with installations' operators allowed to choose between the highest value of the median annual production levels over 2005–08 or 2009–10. $CLEF_{p,t}$ is an allocation reduction factor that is applied to a small minority of products that are not considered to be at risk of carbon leakage. (See EC 2010a.) (These products will see their free allocations reduced by a multiplier of 0.8 in 2013 and the multiplier will decrease linearly each year to reach 0.3 in 2020.) $CSCF_t$ is a uniform, cross-sectoral correction factor that can be applied to ensure that the total free allocation will not exceed the maximum annual amount of free allocation as defined in Article 10a(5) of the ETS directive. It effectively ensures that the level of aggregate free allocation to non-electricity sectors does not rise as a percentage of the total emissions cap over time.

Some additional complexities, including the treatment of cross-installation-boundary heat flows, waste gas recovery, and electricity consumption, can affect an installation's free allocation level. Each of these effects will be mentioned where it is relevant in the discussion below.

4.2 The empirical evidence on Phase III allocations: Some welfare and efficiency implications of the new rules

Data description
This analysis uses EU ETS installations compliance data from the European Union Transaction Log (EUTL) for the period 2008–2011. These data were matched with the preliminary annual free allocation data for each installation for the period 2013–2020 as reported in the National Implementation Measures (NIM) of twenty member states using installation identifiers that were common to both the NIMs publications and the EUTL data on emissions and historical allocations. The NIMs

describe the preliminary free allocation proposed to each installation during Phase III before validation by the European Commission and the application of the cross-sectoral correction factor. Each installation was then further matched with a four-level NACE (Nomenclature of Economic Activities) code to identify its primary production activity. This matching was done using a list of EU ETS installations and their NACE codes provided on the website of the European Commission's Directorate-General for Competition. The missing member states are Belgium, Hungary, Malta, Lithuania, Slovenia, the Czech Republic, and Latvia, whose NIMs were not available or not in a matchable form at the time of writing.

Since the changes in allocations to new entrants in Phase III could not be calculated, we ignores the effects of the benchmarking rules on new entrants in this chapter. Further excluding new entrants, the aviation sector, installations that had left the EU ETS in Phase III, and installations that could not be matched with either a EUTL installation code or a NACE code left a sample of 7,149 installations, which together accounted for 1.46 billion tonnes of CO_2 or approximately 80 percent of EU ETS emissions in 2010 (EUTL 2011). Of these, 4,174 installations were identified by their NACE codes as non-electricity installations and thus subject directly to benchmarking. From those 4,174 installations, 329 specializing in the chemicals and non-ferrous metals sectors were excluded from the analysis because these sectors had their EU ETS perimeter change significantly between Phase II and Phase III and hence changes in allocation could not be attributed to benchmarking alone.

For benchmarked sectors, Phase III free allocations will decrease significantly relative to Phase II
Perhaps the most striking feature of the collected data on allocations in Phase III is that they show that free allocations to benchmarked sectors will decrease significantly in Phase III. For our sample of more than 4,000 benchmarked installations, the aggregate decrease in free allocation relative to Phase II will be 20.6 percent on average. For installations with at least 90 percent of their products considered exposed to the risk of carbon leakage (3,102 of the 4,174 installations in the data sample), the decrease is slightly smaller: 17.6 percent. Note that this is the decrease in allocation *before* taking account of the application of the uniform linear adjustment factor referred to in equation 1. The adjustment factor, which was yet to be announced at the time of writing, has

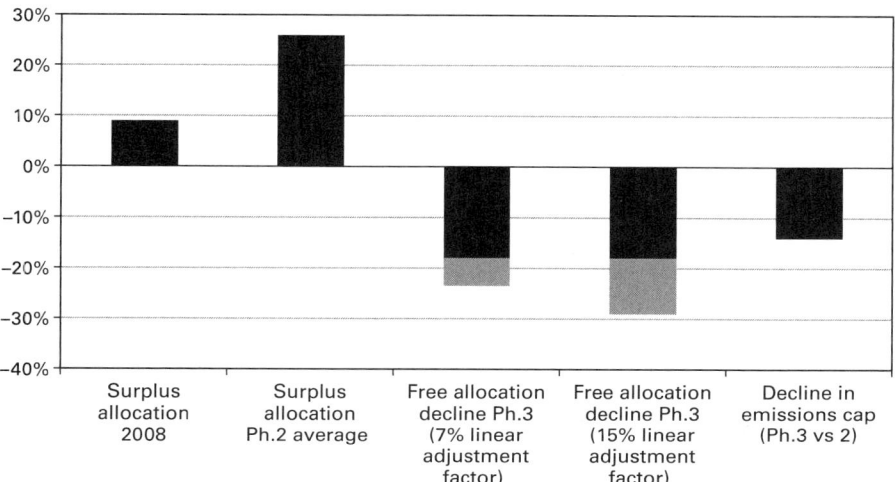

Figure 4.1
Phase II surplus allocations vs. Phase III allocation decreases. The following sectors are included in this analysis: pulp and paper, steel, coke, refining, cement, lime, ceramics, glass, and ferrous metals production.

been variously estimated at between 7 percent (Lecourt 2013) and 15 percent (Graichen et al. 2013) of the preliminary allocated amount. These estimates would therefore imply a further decrease of 5.6–11.9 percent of Phase II allocation levels on top of the 17.6 percent decrease already mentioned.

Figure 4.1 summarizes these results for the 3,102 "carbon leakage exposed" installations and compares them against the aggregate degree of over-allocation experienced by the same 3,102 installations in 2008 and over Phase II as a whole. The results show that the decrease in allocations from simply moving from grandfathering to benchmarks more than offset the aggregate over-allocation in 2008 (before the effect of the economic downturn was felt). Meanwhile, assuming a linear reduction factor of between 7 percent and 15 percent in addition leads to the total decrease in free allocation in Phase III more of less fully offsetting the degree of over-allocation in Phase II on average (which was due mostly to the economic downturn).

As table 4.1 shows, the decreases in free allocation are also quite uniform across the sectors affected by benchmarking. With the exception of what we define here as "other sectors" (which includes a large number of sub-sectors not deemed exposed to carbon leakage and

Table 4.1
Percentage change in allocation by sector.

	Cokery	Refined petroleum products	Glass	Ceramics and brick	Cement	Lime	Pulp and paper	Iron and steel	Other sectors
Aggregate net position 2008[a]	-16	-1	+9	+35	+11	+15	+22	+29	+13
Aggregate allocation change[b]	-17	-24	-24	-16	-13	-19	-21	-13	-37
Median allocation change[c]	-6	-14	-21	-17	-11	-18	-22	-11	-33
Dispersion of allocation changes[d]	25	26	19	28	9	18	1331	82	99

a. calculated as (Allocation – Emissions)/Emissions in 2008 in sector
b. refers to total aggregate reduction in allocation in sector
c. refers to median reduction in allocation of installations in sector
d. refers to average distance from median allocation reduction of installations in sector

therefore facing a larger reduction factor on average), all the decreases are in a relatively narrow range (between –13 percent and –24 percent) before the linear reduction factor. Some sectors will see their allocation changes affected more or less than others by special rules that have greater effects on their specific sector.[3]

Welfare implications: lower risk of windfall profits, minimal carbon leakage risks, and redistributions within member states

The generalized decrease in free allocations in Phase III relative to Phase II is likely to have several important consequences for both the actual functioning and perceptions of the EU ETS. From a welfare distribution perspective, to the extent that allocations well in excess of emissions are likely to be curtailed, risks of windfall profits occurring in the non-power sectors are likely to be much lower than during Phases I and II.

Of course, even at allocation levels below 100 percent of historical emissions, windfall profits could still potentially occur if, Although mostly allocated for free, carbon prices were still passed through to final consumers in marginal product prices in these sectors. However the empirical evidence to date has yet to confirm that this occurs in the sectors concerned. This would also require that international competition is weak enough to allow domestic EU prices to rise to include the opportunity cost of carbon in the concerned sectors. A more realistic way in which windfall profits could still occur under the new benchmarking rules would be if installations were to produce at levels significantly below the ex-ante historical activity level used to calculate their free allocations. As will be shown in the next section, this remains a significant possibility in some sectors because of the effects of the economic downturn and the linking of allocations to ex-ante historical activity level.

Although free allocations have been reduced significantly, it does not appear to be the case that this will create a significant risk of carbon leakage. This is because the free allocations to the main sectors exposed to the risk of carbon leakage will have sufficiently large allocations to mitigate all but a small minority of their net compliance costs. Figure 4.2 shows this. It is calculated assuming a carbon price of €25 per tonne of CO_2, that installations produce at their historical activity levels throughout Phase III and a cross-sectoral correction factor on NIM allocations of –15 percent. It is also assumed that firms have no ability to pass costs through to consumers, abate emissions, or use cheaper

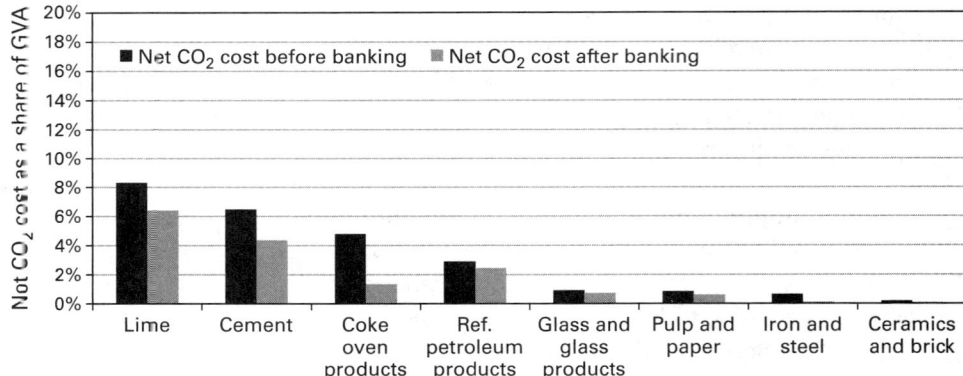

Figure 4.2
Estimated net CO_2 costs for the median installation in each sector as a percentage of sector gross value added at €25 per tonne of CO_2 and with a correction factor of –15 percent. Data on gross value added (GVA) for the EU27 come from the European Commission s Impact Assessment Report of 2009 (EC 2009b). Where a sector has several relevant products with different carbon cost values, the average CO_2 cost impact has been used. Calculations assume that there is no possibility that abatement, carbon price pass-through, or carbon offset use will reduce costs. Indirect carbon costs from electricity prices are ignored. Estimates of banked allowances are based on EUTL data of verified emissions versus Phase II allocations and assume no sales or purchases of allowances.

carbon offsets. Yet even with these assumptions only the lime and cement sectors, for which notoriously high transport costs are a significant barrier to carbon leakage, are estimated to have post-allocation costs of more than 4 percent of gross value added. Moreover, after the effects of banked allowances from Phase II are taken into account the cost estimates are reduced further in every sector. The new allocation rules therefore seem fully capable of mitigating leakage risks in these sectors, even as they reduce free allocations significantly relative to Phase II.

A further interesting distributional implication of the new allocation levels is that they do not appear to strongly redistribute free allowances from certain countries in the EU to others. The redistributive effects of common EU-wide benchmarks are much more important at the intra-member-state level than at the inter-member-state level. Table 4.2 illustrates this point. It decomposes the variance of inter-phase allocation changes in two parts. The first part consists of the variance between each installation and its national average allocation decrease; the second part represents the remaining variance between each installation and the sect oral average allocation change. The results show that

Table 4.2
Decomposition of the installation allocation change variance of benchmarked sectors.

	Pulp and paper	Cokery	Refined petroleum products	Glass	Ceramics and bricks	Cement	Lime	Iron and steel
Inter-country	11%	44%	27%	9%	16%	26%	26%	10%
Intra-country	89%	56%	73%	91%	84%	74%	74%	90%

in the most important sectors some redistribution of allowances will occur across member states, but this redistribution is generally smaller than the redistributions within member states. The use of common, EU-wide emissions performance benchmarks therefore does not appear to induce significant levels of inter-country transfers of allowances relative to EU ETS Phase II. Consequently, the rules do some seem to be disproportionately disadvantaging carbon-intensive industries in certain member states by redistributing allowances away from them to other "better technology."

4.3 Has the potential for allocation-related distortions in the EU's internal market been eliminated by harmonized allocation rules?

It was noted above that one of the primary criticisms of the decentralized free allocations in Phase II was that it allowed for potential competitiveness distortions within the EU's internal market because of differing levels of allocation to installations in the same sector. To what extent has the potential for competitive distortions of the internal market been reduced by moving from a decentralized to a centralized allocation system? Our discussion of this question proceeds in two steps. First, we present an econometric analysis that demonstrates that the more flexible and decentralized approach to free allocations in Phase II led to a significant degree of unexplained heterogeneity in final free allocation rates for installations and member states competing in the same sectors. Second, we show that although these potentially distortionary differences are now eliminated by using harmonized benchmarks, two other important sources of potential market distortions remain: the use of ex-ante output levels and the use of output thresholds to determine allocations.

Modeling approach and data

Phase III's product benchmark-based free allocation rules are intended to provide a greater level of harmonization of the rules for free allocation and the actual levels across installations producing like products. But although Phase III benchmarking rules are harmonized almost by definition, the marginal benefit of this formal harmonization is not immediately obvious. Rather, it depends on the extent to which the earlier system was not harmonized and on the potential for competiveness distortions created by the previous lack of harmonization. The modeling approach developed below attacks this question by first presenting evidence of the extent to which Phase II free allocations could be said to be poorly harmonized both across and within member states. It is shown that the size of the unexplained heterogeneity in allocation levels to installations in Phase II was large enough on a gross-margin basis to have been distortionary to primary product markets in at least one sector. For tractability, the analysis here is restricted to the cement sector. The cement sector was chosen because, of all the sectors, it has the most homogeneous production process as far as emissions are concerned (clinker production). Moreover, to the best of our knowledge, EU ETS cement sector installations do not commonly have significant cross-boundary heat flows, which could meaningfully affect their free allocation levels in Phase III under the new rules.

We begin by specifying the following econometric model of annual allocation changes between Phase II and Phase III in the cement sector:

$$\Delta ALLOC_i = \alpha + \beta_1\, NetPos2008_i + \beta_2 CO2Intensity_i + \sum_{j=2}^{N} \beta_j country_{ij} + u_i. \tag{2}$$

$\Delta ALLOC_{ij}$ is the percentage change in the (average) annual free allocation level of cement installation i in moving from Phase II to Phase III. It is based on the average allocation level per year in Phase II versus that of Phase III. $NetPos08_i$ is a measure for the extent to which installation i was either over-allocated or under-allocated in Phase II. Specifically, it is the ratio of installation i's aggregate free allocation in 2008 divided by its verified emissions in the same year. As above, the year 2008 is chosen to control for the effects of the severe decrease in industrial production in 2009 and thereafter due to economic conditions. Meanwhile, $CO2Intensity_i$ is a measure of the CO_2 intensity of installation i. It is calculated by first identifying the historical activity level year, based on emissions levels in each of the candidate years reported in the EUTL. It is then divided by the average annual free allocation level firms will receive in Phase III (which is the same level each year for these installations, since the linear correction factor is not relevant here). Since the implied production levels of the emissions and the allocations are for the same year for the nominator and the denominator, the result gives a direct measure of the ratio of the actual emissions intensity factor of the installation to the cement benchmark that applies to it. Country dummy variables are included in the regression to make it possible to observe country-specific effects. Here u_i represents variation in allocations due to unobserved factors and α is a constant term.

The logic underlying the model estimation is as follows. To begin with, it is assumed that free allocations to the cement sector in Phase II were based on a pure grandfathering approach—i.e., that free allocations to installations should have depended only on historical emissions, as Ellerman et al. (2010) have suggested they did. If that were true, this system could indeed be considered a well-harmonized free allocation method. After all, such a system would imply that there would be no possibility of systematic over-allocations or under-allocations occurring in some member states but not in others, at least not on a consistent basis if we accept that historical emissions are an unbiased predictor of near-future emissions.

Now, if such a harmonized grandfathering approach was indeed the practice in Phase II, this would imply that, as the EU moved to an EU-wide harmonized benchmark-based allocation, the most important determinant of the size of the decreases in free allocation to each installation must be the difference between historical emissions intensity and the new benchmark of emissions intensity. Thus, in terms of equation 2 only $CO2Intensity_{ij}$ should be statistically significant, and not the other variables.

On the other hand, if, say, Phase II over-allocations (proxied in equation 2 by $NetPos2008_{ij}$) were found to be statistically significant, thus explaining the change in allocations from Phase II to Phase III, this would imply that Phase II allocations were not solely based on historical emissions. Rather, it would mean that some additional unexplained heterogeneous[4] level of allocation was given to different installations during Phase II that was not related to its emissions performance or its historical output level[5] and that this was being "corrected" for by the imposition of the new harmonized EU-wide rules.

Installation-level ordinary least-squares regression results for the cement sector

To estimate the model, we used 223 observations on the cement sector from the already described database. To remove outliers resulting from

Table 4.3
Regression data summary.

	Allocation change	Net position in 2008	CO_2 intensity (actual/benchmark)
Mean	−10.3%	10.9%	94.3%
Median	−11.4%	8.0%	91.9%
S.D.	11.9%	19.4%	11.5%
Max	41.2%	121.6%	153.2%
Min	−52.2%	−32.0%	65.0%
N	223	223	223

capacity changes from biasing the results, the sample was reduced to omit a small number of installations with allocation increases of 50 percent of their maximum and installations which had seen decreases of more than 90 percent in their allocations. Table 4.3 describes various features of the remaining data.

In post-estimation analysis, Shapiro-Wilk tests showed that the estimated residuals were approximately normally distributed, but Breusch-Pagan tests indicated some (weak) evidence of heteroskedasticity. (See appendix below.) Robust standard errors were therefore used for significance testing. A scatter plot showed no signs of outlier-driven bias in the final sample. (See figure 4.3.)

Table 4.4 presents the results of the regression estimation, which was done by the ordinary least-squares (OLS) method. Under all three alternative specifications the $NetPos2008_i$ coefficient was found to be consistently negative and statistically significant at the 99 percent confidence level. The negative sign implies that, on average, the higher (lower) an installation's over- (under-) allocation was relative to its historical emissions, the greater (smaller) that installation's reduction in free allocation was now after it fell under the harmonized rules. These results therefore suggest that a high degree of heterogeneity in free allocations existed in the Phase II NAPs, even after accounting for different levels of allocation due to historical output and emissions intensity of each free allocation.

Moreover, the results strongly indicate that this heterogeneity accounted for a large share of the final Phase II allocation levels to each installation. The coefficient estimates on $CO2Intensity$ and $NetPos2008$ presented in table 4.4 imply that, on average, for every 10 percentage points of decrease in free allocation due to the move to stricter bench-

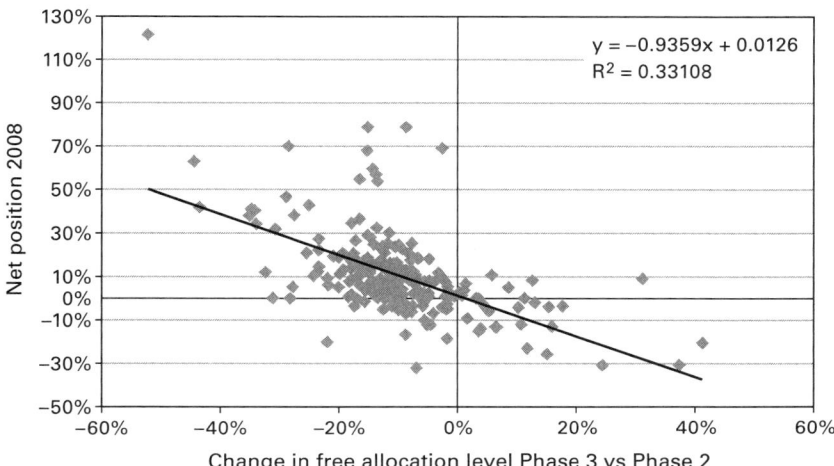

Figure 4.3
Scatter plot of 2008 net positions vs. changes in free allocation changes (cement installations).

marks, roughly 4.5 percentage points are attributable to stricter benchmarks rather than the actual emissions intensity, and approximately 3.5 additional percentage points are effectively correcting for other unexplained sources of differences between the historical emissions level and the historical allocation level in Phase II. The importance of these other unexplained sources of different allocation levels is further highlighted by the fact that they appear to explain approximately one third of the variation in the dependent variable, as seen by the R^2 value of 0.33 in the first specification. This is strong evidence that the lack of formally harmonized allocation rules in Phase II leads to a high degree of variation in allocation levels between installations and that this cannot be explained by their actual emissions levels. It thus provides empirical and quantitative confirmation that concerns raised in the NAP literature are generally justified.

Furthermore, the results do not appear to depend on one or two countries' over-allocating or under-allocating their installations more than others. In specification 4, the *NetPos2008* variable is dropped from the regression and country dummies are added in its place. This specification now captures the average country-specific contribution to the allocation change of each installation, after controlling for its distance from the benchmark level of free allocation and historical output in the

Table 4.4

Installation level regressions results (coefficient estimates for each variable for four different OLS specifications). Standard errors are Newey-West Standard Errors.

Variable	OLS1	OLS2	OLS3	OLS4
NetPos2008	−0.354[a]	−0.376[a]	−0.325[a]	−
CO2Intensity		−0.467[a]	−0.464[a]	−0.400[a]
MS1				−0.123[a]
MS2				−0.077[b]
MS3				−0.166[a]
MS4				−0.312[a]
MS5				−0.172[a]
MS6				−0.137[a]
MS7				−0.118[a]
MS8				−0.150[a]
MS9				−0.190[a]
MS10				−0.200[a]
MS11				−0.156[a]
MS12				−0.131[a]
MS13				−0.341[a]
MS14				−0.180[a]
MS15				−0.250[a]
MS16				−0.265[a]
	None	None	Yes	Yes
	−0.065[a]	0.439[a]	0.383[a]	−0.516[a]
Descriptive statistics				
R^2	0.33	0.53	0.67	0.42
Prob > F	0.000	0.000	0.000	0.000
Observations	223	223	223	223

a. statistically significant at 2.5% level
b. statistically significant at 5% level
c. statistically significant at 10% level

same manner as before. The dummy variable for each member state now captures the country-specific level of allocation in Phase II that is being adjusted for in Phase III, relative to Austria, which is the base case and is represented by the constant term.

The results of specification 4 indicate that all seventeen member-state dummies (including the constant) were found to be statistically significant. The country dummy coefficients differ widely, ranging from −0.077 to −0.341 with a median of −0.17. The standard deviation of the coefficients themselves, 0.073, is also well outside the average 95 percent confidence interval estimate for all of these coefficients, which is ±0.033. These results confirm that the previous result was not driven by one or two countries but was an EU-wide phenomenon.

What is the scope for market distortions in the EU cement sector?
It is instructive to give a quantitative indication of the risk of distortions on the primary product markets with such differing levels of allocations in Phase II could have created. To do so, it is first assumed that only residual CO_2 emissions costs after deducting the free allocation compensation are incorporated into the unit price of cement by producers, in keeping with the idea of a threat of international competition if prices rise to reflect CO_2 opportunity costs. Symmetrically, it is assumed excess free allocations over emissions are considered by firms as a per unit subsidy, i.e.,

Unit CO_2 cost
     = EUA price × (emissions per unit − free allocation per unit).

$$(3)$$

Second, the *NetPos2008* coefficient estimate from regression 3 in table 4.4 is exploited to obtain fitted estimates of the part of the free allocation of each installation (as a share of its emissions) which is due to the unexplained heterogeneity identified earlier. Third, since the econometric analysis in the preceding section was based on 2008 data, 2008 data on the cement sector's emissions per unit of output are obtained from the World Cement Sustainability Initiative GNR database and are assumed to be 0.65 tonnes of CO_2 per tonne of cement. Finally, representative Portland cement prices of €75 and €95 per tonne are used. Table 4.5, which is based on these data, presents five different estimates of the effective subsidy (or cost) rate that are estimated to have been received by the different installations. Each estimate corresponds to a different quintile of the distribution of allocation levels observed in 2008 and provides the associated subsidy (or cost) level at carbon prices of €25 per tonne of CO_2 (average 2008 prices). The size of the subsidy relative to the two different possible assumptions about output prices are provided to give an indication of the possibility for distortions. The results show that the size of the subsidies in the zeroth and first quintiles are significant as a share of product prices, particularly in the low-demand scenario. Note that, although the effective subsidy as a percentage of the final product price is only 2–3 percent

Table 4.5
Effective per-unit subsidy rates to cement installations due to free allocation in Phase II (by quintile).

Quintile	Per-unit subsidy (€ per ton of cement)	% of output price (low demand)	% of output price (high demand)
5th	–3.95	–5.3%	–4.2%
4th	0.37	0.5%	0.4%
3rd	0.96	1.3%	1.0%
2nd	1.38	1.8%	1.5%
1st	2.05	2.7%	2.2%
0th	11.28	15.0%	11.9%

for the first quintile, 2–3 percent of final product prices translate to a much higher share of net margins. Moreover, there is a particularly wide range of effective subsidy rates between the zeroth, first, and fifth quintiles—particularly in the low-demand scenario, with the "effective free allocation subsidy" ranging from –5.3 percent to 15 percent of output prices. This estimation from the cement sector seem to indicate potential for primary market distortions exist and that the move to Phase III benchmarks are an important development, at least for sectors with cost structures similar to those of the cement sector.

Ex-ante output-data distortions: the cases of Greece, Italy, Ireland, Spain, and Portugal

The discussion on the potential for distortions in EU ETS Phase II has thus far ignored an important caveat: Although under benchmarking the product benchmarks applied to the allocation given to each installation are the same, the historical output or activity levels (*HAL* in equation 1) are not the same for every installation. For that matter, so will be capacity utilization rate, and thus the level of production relative to the *HAL* will be different.

Figure 4.4 shows two comparisons using the example of Greek, Italian, Irish, Spanish, and Portuguese cement installations in the EU ETS. First, it compares the 2012 emissions and 2013 allocations of these installations (including the application of a 7 percent cross-sectoral correction factor). It is clear that if emissions levels in 2013 are similar to those of 2012 (not an unreasonable assumption) the majority of installations would receive a very large over-allocation of allowances relative to their emissions. The main reason for this is that 2013 alloca-

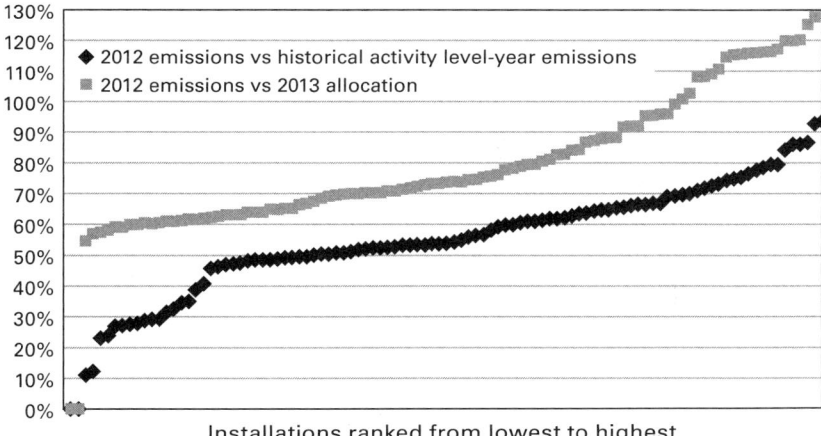

Figure 4.4
2012 emissions vs. 2013 allocations and historical emissions in Greek, Italian, Irish, Spanish, and Portuguese cement sectors.

tions are based on historical output levels that are related to periods before the global financial crisis and the severe collapse in demand that ensued. This is shown by the second data series presented in figure 4.4, which shows the ratio of the emissions of these installations in 2012 to their emissions in the historical activity level year to determine its Phase III free allocations.

On the basis of the analysis of effective subsidy rates presented earlier in this chapter, it is clear that significant potential still remains for internal EU-market distortions to arise in the cement or similar sectors because of these very different levels of emissions relative to free allocations. The combination of a formula based on distant ex-ante production levels and a severe sectoral downturn have created similar-sized risks of distortions in the internal market for cement and clinker.

Conclusion

The preceding discussion has shown that Phase III allocation rules have a number of desirable distributional welfare and efficiency characteristics relative to those of Phase II allocations. Most interesting, our analysis has revealed that Phase II free allocations were not perfectly harmonized across member states, and that some member states' allocations in the cement sector exhibited systematic country-specific allocation differences between countries. These differences were economically significant. The use of harmonized benchmarks has now been corrected.

Evidence strongly suggests that the introduction of benchmarking does not lead to significant redistributions of allowances from industries in certain member states to industries in others. This is also an interesting property of a free allocation mechanism in the European context.

On the other hand, although harmonization of allocation rules has removed one source of potential competitiveness distortions identified by the NAP literature (different allocation rules in different member states), other kinds of competitiveness distortions are still possible owing to the use of ex-ante output data to determine allocations in a time of low capacity utilization. We therefore conclude that the benchmarking system is necessary but not a sufficient condition for a distortion-free anti-leakage mechanism in the ETS.

Appendix: Results of pre-estimation econometric tests

The results of pre-estimation econometric tests are given in tables 4.6 and 4.7.

Table 4.6
Shapiro-Wilk test for normally distributed residuals. H_0: Residuals are normally distributed. Conclusion: Insufficient evidence to reject H_0.

Variable	Observations	w	v	z	Prob > z
Residuals	223	0.97053	1.187	0.362	0.35875

Table 4.7
Breusch-Pagan test for heteroskedasticity. H_0: $\sigma(i)^2 = \sigma^2$ for all i (i.e., there is no heteroskedasticity in errors). Conclusion: Insufficient evidence to reject H_0.

$X^2(3)$	5.14
Prob > $X^2(3)$	0.1617

Notes

1. In some cases benchmarks for specific products cannot easily or practically be used and so hierarchy of fallback approaches is used, based first on heat and then fuel consumption benchmarks and, if these are not possible, historical process emissions multiplied by 0.97 are used.

2. Where the best 10 percent of installations emissions intensity could not be gauged, fallback approaches based on best available technology literature were used.

3. For example, despite generally large allocations in 2008 relative to emissions, the steel sector sees a reduction of only 13 percent on aggregate in Phase III. This reflects the fact that, under benchmarking, a significant share of emissions allowances is allocated to steel sector installations for waste gas emissions that occur offsite. The cost of the emissions from these waste gases however affect the electricity price steel makers receive for the waste gases that are provided to electricity generators, who provide electricity in return. Similarly, the refined petroleum sector sees a larger decrease (24 percent) despite its deficit in 2008 because of its large share of in-house generation of electricity, which is not compensated via the benchmarking mechanism but via other mechanisms, such as direct state aid.

4. If the additional free allocation were homogeneous across member states, it would be captured by the constant and sector dummies.

5. Because the dependent variable is based on the historical activity level used in the benchmark calculation, historical activity is controlled for implicitly in the model.

References

Abrell, J., A. Ndoye, and G. Zachmann. 2011. *Assessing the Impact of the EU ETS Using Firm Level Data*. Brueghel.

Åhman, M., D. Burtraw, J. Kruger, and L. Zetterberg. 2007. A ten-year rule to guide the allocation of EU emission allowances. *Energy Policy* 35 (3): 1718–1730.

Anderson, Barry, and Corrado Di Maria. 2011. Abatement and allocation in the pilot phase of the EU ETS. *Environmental and Resource Economics* 48 (1): 83–103.

Benz, E., A. Löschel, and B. Sturm. 2009. Auctioning of CO_2 emission allowances in Phase 3 of the EU Emissions Trading Scheme. *Climate Policy* 10 (6): 705–718.

Betz, R., W. Eichhammer, and J. Schleich. 2004. Designing national allocation plans for EU emissions trading: A first analysis of the outcomes. *Energy & Environment* 15 (3): 375–425.

Betz, R., K. Rogge, and J. Schleich. 2006. EU emissions trading: An early analysis of national allocation plans for 2008–2012. *Climate Policy* 6 (4): 361–394.

Betz, R., and M. Sato. 2006. Emissions trading: Lessons learnt from the 1st phase of the EU ETS and prospects for the 2nd phase. *Climate Policy* 6 (4): 351–359.

Böhringer, C., C. Fischer, and K. E. Rosendahl. 2010. The global effects of subglobal climate policies. *B.E. Journal of Economic Analysis & Policy* 10 (2) (Symposium), article 13.

Bovenberg, A. L., and L. H. Goulder. 2002. Environmental taxation and regulation. In *Handbook of Public Economics*. 1st ed. vol. 3. ed. A. Auerbach and M. Feldstein. Elsevier.

Clò, S. 2010. Grandfathering, auctioning and carbon leakage: Assessing the inconsistencies of the new ETS Directive. *Energy Policy* 38 (5): 2420–2430.

Crampton, P., and S. Kerr. 2002. Tradeable permit auctions: How and why to auction not grandfather. *Energy Policy* 30: 333–345.

del Rio Gonzalez, P. 2006. Harmonization versus decentralization in the EU ETS: An economic analysis. *Climate Policy* 6: 457–475.

Demailly, D., and P. Quirion. 2006. CO_2 abatement, competitiveness and leakage in the European cement industry under the EU ETS: Grandfathering versus output-based allocation. *Climate Policy* 6: 93–113.

Dröge, S., and S. Cooper. 2010. *Tackling Leakage in a World of Unequal Carbon Prices*. Climate Strategies.

EC (European Commission). 2003. Determining, pursuant to Directive 2003/87/EC of the European Parliament and of the Council, a list of sectors and subsectors which are deemed to be exposed to a significant risk of carbon leakage (2010/2/EU).

EC. 2009a. Directive 2003/87/EC of the European Parliament and of the Council of 13 October 2003 establishing a scheme for greenhouse gas emission allowance trading within the Community (consolidated version).

EC. 2009b. Impact Assessment Accompanying document to the Commission Decision determining a list of sectors and subsectors which are deemed to be exposed to a significant risk of carbon leakage pursuant to Article 10a (13) of Directive 2003/87/EC, SEC(2009) 1710 final.

EC. 2010a. Decision 2010/2/EU determining, pursuant to Directive 2003/87/EC of the European Parliament and of the Council, a list of sectors and subsectors which are deemed to be exposed to a significant risk of carbon leakage.

EC. 2011. Decision 2011/278/EU on determining Union-wide rules for harmonized free emission allowances pursuant to Article 10a of Directive 2003/87/EC.

Ellerman, A. D., F. Convery, and C. De Perthuis. 2010. *Pricing Carbon: The European Union Emissions Trading Scheme*. Cambridge University Press.

Ellerman, A. D., B. Buchner, and C. Carraro, eds. 2007. *Allocation in the European Union Emissions Trading Scheme: Rights, Rents, and Fairness*. Cambridge University Press.

EUTL. 2011. European Union Transaction Log (http://ec.europa.eu/environment/ets/).

Golombek, R., S. A. C. Kittelsen, and K. E. Rosendahl. 2013. Price and welfare effects of emission quota allocation. *Energy Economics* 36:568–580.

Graichen, V., K. Schumacher, S. Healy, H. Hermann, R. Harthan, M. Stork, B. Borkent, A. Mulder, P. Blinde, and L. Lam. 2013. *Support to the Commission for the determination of the list of sectors and subsectors deemed to be exposed to a significant risk of carbon leakage for the years 2015–2019 (EU Emissions Trading System)*. European Commission.

Kruger, Joseph, Wallace E. Oates, and William A. Pizer. 2007. Decentralization in the EU emissions trading scheme and lessons for global policy. *Review of Environmental Economics and Policy* 1 (1): 112–133.

Lecourt, S. 2013. EU ETS Phase 3 benchmarks: Implications and potential flaws. Working paper 2013-05, Cahiers de la Chaire Economie du Climat.

Martin, R., M. Muûls, L. B. de Preux, and U. J. Wagner. 2012. Industry compensation under relocation risk: A firm-level analysis of the EU emissions trading scheme. Working paper RSCAS 2012/37, Robert Schuman Centre for Advanced Studies.

Monjon, S., and P. Quirion. 2010. How to design a border adjustment for the European Union Emissions Trading System. *Energy Policy, Elsevier* 38 (9): 5199–5207.

Montgomery, W. D. 1972. Markets in licenses and efficient pollution control programs. *Journal of Economic Theory* 5 (3): 395–418.

Neuhoff, K., M. Ferrario F. Grubb, E. Gabel, and K. Keats. 2006. Emission projections 2008–2012 versus national allocation plans II. *Climate Policy* 6 (4): 395–410.

Pahle, M., L. Fan, and W.P. Schill. 2011 How emission certificate allocations distort fossil investments: The German example. *Energy Policy* 39 (4): 1975–1987.

Palmer, K., D. Burtraw, and D. Kahn. 2006. Simple rules for targeting CO_2 allowance allocations to compensate firms. *Climate Policy* 6: 477–493.

Pearson, A. 2010. *The carbon rich list: The companies profiting from the EU Emissions Trading Scheme*. Sandbag Climate Campaign.

Sartor, O. 2012. Carbon leakage in the primary aluminium sector: What evidence after 6.5 years of the EU ETS? Working paper, CDC Climat, Paris.

Sijm, J., K. Neuhoff, and Y. Chen. 2006. CO_2 cost pass-through and windfall profits in the power sector. *Climate Policy* 6 (1): 49–72.

Smale, R., M. Hartley, C. Hepburn, J. Ward, and M. Grubb. 2006. The impact of CO_2 emissions trading on firm profits and market prices. *Climate Policy* 6 (1): 31–48.

Trotignon, R., and A. Delbosc. 2008. *Allowance trading patterns during the EU ETS trial period: What does the CITL reveal? Climate Report 13*. CDC Climat.

5 The Influence of Permit-Price Uncertainty and Lobbying on Energy Investments

Philipp Hieronymi and David Schüller

The importance of environmental issues in the policy arena has changed substantially over the last decades. Issues such as acid rain, the depletion of the ozone layer and climate change have put environmental concerns high on the priority list of policy makers. Among the regulatory approaches to dealing with these concerns are market-based approaches, voluntary agreements, and command-and-control approaches. Not surprisingly, these regulatory approaches have affected the energy, oil, and automobile industries more than other industries and have led to the formation of lobbying groups that try to influence the political process in their favor. With the introduction of mechanisms supporting renewable energy, especially in many European countries, producers of renewable energy have also become an important lobbying force over the last decade. (One example is the European Wind Energy Association, which has more than 700 members from 60 countries.[1]). Furthermore, non-governmental organizations (NGOs), usually with views opposing those of the industrial lobbies, have increasingly become involved in the political process.

Early political economy investigations of the effect of lobbying on political decision making focused on the losses caused by competing lobbying groups (Becker 1983, 1985). Grossman and Helpman (1994) presented a model in which several groups made political contributions in order to influence trade policy. More recently, Conconi (2003) considered how "green lobbies" affect international trade. Furthermore, a literature related to lobbying and its effect on emission markets arose. Lai (2006) focused on how lobbying affects the initial allocation of pollution rights. Lai (2008) investigated whether auctioning, grandfathering, or a hybrid instrument was chosen when environmental and industrial groups could influence the process. Hanley and Mackenzie (2010) introduced a three-stage model in which firms had the

opportunity to increase their own permit allocation and permit allocation in general. Habla and Winkler (2013) introduced a model in which
governments first decided whether to become part of an international
permit market and in the second stage about national permit allocations. In both of Habla and Winkler's stages, lobbying groups were able
to influence the outcome.

The above-mentioned literature is missing the element of uncertainty. When one is considering how uncertainty affects investors in
the electricity sector, where investments are usually irreversible, the
methodology of real options is a useful approach. Relative to methodologically simpler approaches, such as net present value, a real-options
approach takes future realizations into account, allowing optimal
timing of the investment decision. After the original development of
financial options valuation by Black and Scholes (1973) and Merton
(1973), their techniques were adopted for real investments in the physical sense by Myers (1977). Good overviews with numerous examples
were published by Dixit and Pindyck (1994) and Trigeorgis (1996).
Applications of a real-options approach to the environmental and
energy questions have been on the rise in recent years,[2] but none of the
authors who have applied that approach have considered the effect of
lobbying.

In this chapter, attempting to bring the fields of real-options analysis
and political economy closer together, we address the following
questions:

• How does lobbying for wind or gas power affect investment
decisions under permit-price uncertainty?
• When is the option to lobby used?
• Is the wind or gas lobbying option used?

These questions are relevant because large energy companies seem
reluctant to invest large sums in renewable energy despite a range of
newly provided incentives, such as feed-in tariffs. Furthermore, large
energy companies do not seem to engage in lobbying to enhance the
profitability of renewables. We want to consider whether this behavior
can be explained in an uncertainty framework in which lobbying for
fossil energy implies a reduction in the investment cost, and lobbying
for wind energy implies a premium on the price of electricity relative
to fossil energy. Also, we want to get a sense of the magnitude of the
incentives that are necessary to affect the investment behavior of large
energy companies. The results we generate with our model are best

interpreted as information for a government that wants to evaluate the effectiveness of its renewable policy while allowing lobbying for fossil energy, not as an indication of actual investment decisions of companies. A much more complex model including daily data and a range of other stochastic processes would be required to simulate this.

In order to answer the questions posed above, we incorporate uncertainty in a simplified setting where an investor can choose between investing in a wind or a gas power plant over a fixed time horizon. Uncertainty stems from the stochastically evolving permit price, which also affects the electricity price. The permit price follows a geometric Brownian motion with a positive trend. In the first year 40 gas power plants are installed, and each year one plant must be replaced. Simultaneously, the investor can choose to lobby the government to decrease the permit price for one period by a fixed percentage or to receive a higher remuneration for wind energy. Lobbying expenses are irreversible once incurred and cannot be used for other purposes. The opportunity to spend money on lobbying efforts therefore represents a real option to render gas or wind power plants more profitable in a certain period. By employing this approach, the investor takes into consideration the complete time horizon when making his decision, which might lead to a postponement of lobbying efforts to later periods when the price is likely to be higher.

5.1 Lobbying and the politics of climate change

The main interest groups involved in climate-change policies in the European Union and the United States are the democratic state, polluting industries, and non-governmental organizations (Svendsen 1999; Gullberg 2008). Fredriksson et al. (2005) show that in more democratic states environmental policy is more stringent, but that a threshold must be reached before this effect plays a role. Furthermore, they show that when this threshold has been passed the number of environmental NGOs also affects the stringency of environmental policy. Though NGOs have been able to have been able to exert more and more influence (Gulbrandsen and Andresen 2004), energy-intensive companies and polluting industries have considerably more resources with which to influence the process and are better organized (Gullberg 2008; Wettestad 2009). In the European Union, large energy producers with high CO_2 levels have a sector association called EURELECTRIC. A comprehensive overview of lobbying organizations for different industries can be found in Markussen and Svendsen 2005. In some cases (for example, when pushing for wind-energy subsidies), environmentalists and industrialists cooperate (Brandt and Svendsen 2004). However, usually they vindicate starkly contrasting views. A case study by Skodvin et al. (2010) shows how the option of introducing a large-scale auctioning system, favored by NGOs over the grandfathering approach, was discarded as a result of lobbying by polluting industries. In some cases, even industrial lobbying groups have differing goals. Helm (2010) looks at the British renewable and emissions policy, and also finds evidence of lobbying activity that influenced the policy-making process in such a way as to make it friendlier to industry.

Whereas some energy lobbying groups (mostly carbon-intensive producers that rely on coal energy) favor a low price for CO_2 permits, others (such as producers that own many gas and nuclear power plants) favor a high permit price. The reason for this apparently odd behavior by gas and nuclear power producers lies in the way rents accrue in the energy market with the introduction of CO_2 permits (Keppler and Cruciani 2010). The price of electricity at any time is set by the power plant with the highest-cost base load, usually an emission-intensive fossil-fuel plant. So-called infra-marginal rents accrue when a high-cost power plant is producing at the same time as a lower-cost power plant that can sell its electricity for the same price. In the past these infra-marginal rents compensated producers for the various types of production and dispatch risks. The distribution and significance of these infra-marginal rents has changed with the introduction of CO_2 permits. In most cases the permit price was fully passed on to the consumer, even though permits were allocated via grandfathering. Carbon-intensive producers were able to reap some extra profit, while relatively carbon-free energy production from gas and nuclear power became significantly more profitable the higher the permit price (Keppler and Cruciani 2010). These effects are also likely to occur when permits are fully auctioned, favoring low-emission energy plants even further.

In the following model description, we abstract from this situation and focus on what happens when higher permit prices would actually represent a cost increase instead of an net increase in profit due to infra-marginal rents. Furthermore, we do not consider fossil fuel production competition due to already operating nuclear and coal plants, which lies at the core of the above described behavior. Instead, we investigate the incentive to invest in renewable energy rather than fossil-fuel power, whereas lobbying can influence the profitability of either renewable or fossil fuels.

5.2 The model

Profit function and investment option

In our model we look at an energy investor that represents the energy industry as a whole. We consider this as the appropriate modeling approach, since lobbying in disaggregated form on energy policies on a company by company level is usually not observed. Furthermore, one company alone is not able to influence the permit price, and therefore each single company is a price taker. However, when all the major companies are affiliated with an international lobbying organization, the permit price can be influenced. We abstract from cartel issues and do not consider tacit collusion and price manipulation. Instead, we consider the possibility of influencing the political process in order to reduce the price of emissions permits in order to render investments in fossil energy more profitable, or to engage in lobbying activity that leads to a higher price for electricity generated by wind energy. Our model simulates a 40-year period. In every year, a predetermined amount of existing infrastructure has to be replaced. The investor can choose to replace the existing infrastructure with either a wind power or a gas power plant. Both types of investments are considered irreversible. Companies have to comply with emission-abatement requirements when investing in gas power plants, which is done by obtaining emission permits. Permits are not grandfathered but have to be purchased. An increase in the permit price triggers an increase in the price of electricity. How much of the cost can be passed through is fixed at the beginning of the model.

The investor decides each period whether to invest in costly and irreversible lobbying efforts. By employing a real-options methodology, the investor can postpone this investment, taking future realizations into account. This increases the value of the lobbying options relative to a net-present-value approach. If lobbying is carried out, the permit price is reduced by a fixed proportion between 0 and 1 or the electricity price received for energy from wind power plants is increased by a fixed amount[3]. As a simplifying assumption, reductions in the permit price or increases in the price of energy due to lobbying are intertemporally unrelated. This means that increases or reductions have to be renegotiated each period and there is no permanent

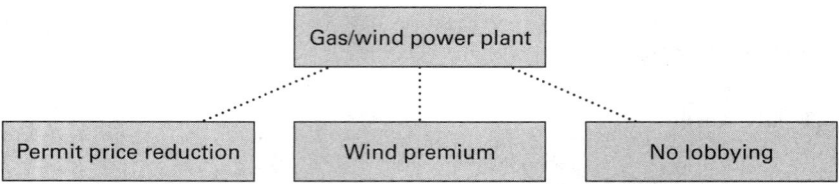

Figure 5.1
Investment/lobbying decision at time *t*.

effect. Both permits and energy are set back to their respective prices without the influence of lobbying efforts at the end of a period. The decision tree an investor faces each period is illustrated in figure 5.1.

The motivation for the reduction of permit prices stems from the negotiations concerning the long-term CO_2 reduction target in the EU. In recent negotiations, the reduction target until 2020 was fixed at 20 percent of 1990 emissions levels, rather than at 30 percent. The long-term reduction goal until 2050 is still under discussion. As the overall cap for the ETS is decided by the reduction goal, lobbying on this issue is most closely related to our lobbying approach. If the reduction target can successfully be pushed down via lobbying, the permit price will decrease because the cap will be at a higher level. Concerning the wind-energy premium, feed-in tariffs for renewable energy production are most closely related to our modeling approach. Feed-in tariffs pay producers of renewable energy a premium over the market price, usually fixed for a time period exceeding 10 years. These premia have become an important tool for the rapid development of a renewable energy capacity. The profit function is defined as follows:

$$\pi(x_t, a_t, b_t, P_t^P)$$
$$= q^e P^e w_t(b_t) - q^g(x_t)P^g - OC(x_t) - c(a_t) - l(b_t) - (1 - L(b_t))q^P(x_t)P_t^P, \tag{1}$$

where

x_t = State the system is currently in at time t; gas plants and/or wind power plants operating,
a_t = Energy investment chosen at point t; gas plant or wind power plant,
b_t Lobbying choice at point t; gas or wind lobbying or no lobbying,
$w_t(b_t) \in [0, 30]$ wind premium in period t,

q_e, P_e = Quantity and price of electricity,
q_g, P_g = Quantity and price of gas,
OC = Operating and maintenance cost,
$c(a_t)$ = One-time investment cost depending on the action chosen,
$l(b_t)$ = One-time lobbying cost,
$L(b_t) \in [0, 1]$ = Permit-price reduction when lobbying for gas power is
carried out,

and

q^p, P^P = Price and quantity of permits purchased.

The time horizon of the model is 40 years. This time horizon has been chosen because the average operating life of a gas power plant is about 30 years. The average operating life of an onshore wind power plant is 25 years (IEA 2010).

Permit-price process
The price for emission permits follows a stochastic process. In accordance with results from the current literature that model the price development for European Emissions Trading System permits (Yang et al. 2008; Fuss et al. 2008, 2009, 2010), we assume a geometric Brownian motion with a 5 percent positive trend. This is essentially a political assumption. If the global goal is to limit the climate-change-driven temperature change to 2°C, and if permit markets are to play an important role in this, permit prices must increase with a 5 percent trend. Otherwise the incentives to invest in carbon-free forms of energy are insufficient. A second way of interpreting this modeling approach is that the 2°C goal with the necessary permit-price increase represents the constant lobbying pressure by NGOs working for a sustainable future.

The price process can, thus, be described as follows: Let $(\Omega, \mathcal{F}, \mathbb{P})$ be a probability space and let W_t be a Wiener process with respect to a probability measure $\mathbb{P}$. Let $(\mathcal{F}_t)_t$ be the filtration generated by W_t. The price process P_t is described by the equation

$$\frac{dP_t}{P_t} = \mu dt + \sigma dW_t. \tag{2}$$

Hence P_t is a stochastic process following a geometric Brownian motion. Here $P_{1,t}$ models the price process for EUA permits, μ is the drift parameter, and σ is the volatility parameter.

Bellman equation

The investor knows that the prices of emissions permits evolve stochastically from a known starting value. Therefore, he has to decide what type of power plant to invest in and whether to invest in lobbying. In order to analyze this, we use the following Bellman equation, solved recursively:

$$V_t(x_t, c_t, b_t, P_t^P) = \max_{a_t \in A_t(x_t), b_t \in B_t(x_t)} \{\pi(x_t, a_t, b_t, P_t^P)$$

$$+ e^{-r} E[V_{t+1}(x_{t+1}, P_{t+1}^P,) \mid P_t^P]\}. \tag{3}$$

The first part of the equation is the immediate revenue stream a company obtains when investing in either of the two power plants or in both at the same time. $A_t(x_t)$ is the set of feasible actions for a given state x_t and r. For example, if a gas plant has already been built, $A_t(x_t)$ comprises two possible actions: building a gas power plant and building a wind power plant. $B_t(x_t)$ stands for three possible lobbying actions: wind premium, permit-price reduction, no lobbying.

5.3 Data

Power plant data

The costs and CO_2 emissions listed in table 5.1 are derived from the 2010 edition of *Projected Costs of Generating Electricity* (IEA 2010) and current market data from the European Energy Exchange (EEX 2014).[4] We compare a combined-cycle gas turbine (CCGT) with an onshore wind power plant. We follow Fuss et al. (2010) in assuming that opera-

Table 5.1
Costs of generating electricity.

	CCGT	Onshore wind
Generation capacity (MW)	480	45
Capital cost (overnight cost) (US$/kW)	1,068.00	2,348.00
Fuel cost (US$/MWh)	36,45	0
O&M (US$/MWh)	4.48	21.92
CO_2 emissions (tCO_2/MWh)	0,33	0
Load factor	85%	26%
Electricity price (US$)	81	81
Lifetime	30 years	25 years
Risk-adjusted rate	5%	5%

sources: IEA 2010; EEX 2014

tion and maintenance cost (O&M) and capital cost are deterministic. The load factor is the average yearly usage of the full capacity of the power plant. Capital cost can be considered as overnight cost, which are the cost that would apply if the plant could be constructed overnight.[5] Using overnight costs for two different types of plants is a valid assumption if the lead time does not differ substantially, as is the case for wind and gas power (IEA 2010; Kettunen et al. 2011). The electricity price is set at €60, which lies approximately in between the medium-term baseload and the peak market price (EEX 2014).[6] Concerning the deterministic gas price, long-term contracts are common for gas deliveries and current research estimates a stable mean-reverting trend for gas prices (Abadie and Chamorro 2006; Boogert and de Jong 2011). We do not consider technological progress because both power-plant types are mature technologies that, we assume, experience similar speed of technological improvement in the long term, which leaves the overall investment decision unaffected. In the model, we consider the replacement of a 400-megawatt plant each year, which is roughly equal to the production capacity of one average gas power plant or 34 wind turbines. This implies that investment behavior is driven by the relative cost differences of these power-plant types, not by the difference in absolute investment size between an average gas power plant and a wind power plant

We assume a risk-adjusted rate of 5 percent. These values are in accordance with the literature on power investment (Kaplan 2008; IEA 2010; Kettunen et al. 2011). We use the same risk-adjusted rates for both power-plant types, since the risk-adjusted rate only matters for operation and construction risk which is similar for both investment types.

Permit prices, electricity price, and volatility
In table 5.2 the assumptions concerning the permit-price process are listed.[7] We take current future prices of EU ETS as a point of departure.[8] In accordance with current literature and modeling, we assume that the trend of the EU ETS price is 5 percent and the volatility is 20 percent (Fuss et al. 2010; Kettunen et al. 2011). We emphasize again that the 5 percent increase is a political assumption. If permit markets are to play a dominant role in offering the necessary incentives to limit the temperature increase driven by climate change to 2°C by the end of the century, such a price increase will be necessary (Fuss et al. 2010). An alternative interpretation of this assumption is that this is the result of

Table 5.2
Assumptions regarding electricity and permit prices.

	EU ETS
Permit-price trend	5%
Permit-price volatility	20%
Permit starting price	US$8.80
Electricity price pass-through	30%
Wind-power premium	US$27.00

environmental pressure groups influencing the political process, thereby representing the counter weight to the industrial lobbying efforts. The high volatility found in the literature is due to the strong influence of political decision making on this market, which changes frequently. The set-aside currently being debated in order to revive the dwindling CO_2 price would lead to a surge in permit prices.[9] Fezzi and Bunn (2009) estimate that the pass-through rate for the EU ETS is around 30 percent, so we use that as a baseline value. Renewable energy currently receives a range of support besides the indirect support of putting a price on CO_2 emissions. We choose to incorporate these various form of support via a price premium for wind energy on the electricity price. For the size of the subsidy, we took the German market as an example. In the German case, onshore wind receives initially a very high premium, which, according to a Wikipedia article consulted in 2014, can last up to 13 years. The premium is then slowly reduced to match market prices. On the basis of the current feed-in tariff structure, onshore wind energy then receives a premium of approximately €20 per megawatt-hour. In the context of our lobbying model, this premium should be interpreted as the amount that lobbying is able to convince policy makers is appropriate. Put differently, it shows the receptiveness of the political system to lobbying efforts.

Lobbying cost
In the United States, expenditures related to lobbying efforts have to be made public and can be viewed on line.[10] For example, the lobbying expenditure of electric utilities was about US$145 million in 2011. Not included in this number are donations to political parties. Taking this number as an indicative lower boundary for lobbying expenditures for both wind and gas lobbying, we perform sensitivity checks with respect to different cost levels. The cost of lobbying enters the model as a per-

centage of investment cost of a gas power plant, as described above. The investment costs of a gas power plant, US$427 million, are based on the data provided above.[11]

As of 2012, about 480 gas power plants were operating in Europe, with an average capacity of about 360 MW.[12] In 2010, there were 1,600 gas plants,[13] with a generating capacity of 1,000 gigawatts,[14] in the United States—roughly four times as many plants, with three times the generating capacity, as in Europe. Assuming that the larger the market the greater the expense of lobbying, we estimate that the lobbying expenditures in Europe should be around one fourth of those in the US. This implies an indicative lower bound of about US$36 million. Setting this in relation to the investment cost for a gas power plant, the lower bound for lobbying expenses is around 8.5 percent of the investment cost in our model. Since we only consider a model with 40 plants operating, this can be considered a conservative estimate, since there are nine times as many plants operating in Europe, which implies that much more capacity has to be replaced than in our model. Therefore, the actual lobbying expenditure per power plant is probably smaller.

5.4 Results

Each scenario was run 2,000 times. The first graph for each scenario shows the number of gas plants installed in a certain year. The model starts with forty gas plants, one of which has to be replaced every year. Thus, if fewer than forty plants are gas plants, the remaining ones are wind power based. The dotted line in figure 5.2 shows the number of gas plants installed when companies can lobby; the solid line shows the case when no lobbying is possible.

First we consider the effect of allowing lobbying for wind relative to the case where only gas lobbying can take place. Unless reported differently, certain values remain the same for all scenarios in this section and will not be reported again: The volatility rate of the permit price was set at 20 percent, the emissions price trend at 5 percent, the pass-through rate at 30 percent, and the interest rate at 5 percent. In figure 5.2 the results for the case in which only gas lobbying is allowed are reported. When only gas lobbying is allowed, gas plants are replaced by gas plants and only toward the end of the time horizon does wind investment become a possible choice, as can be seen in figure 5.3. Figure 5.4 shows lobbying activity. No lobbying for gas power is carried out

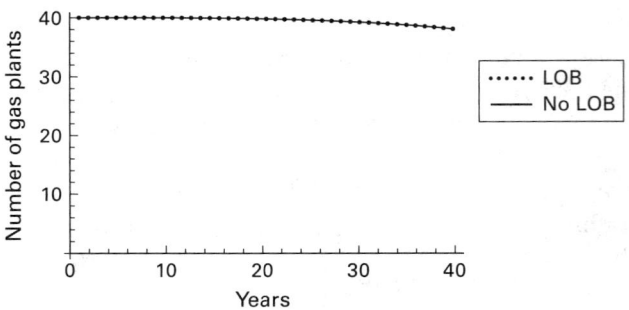

Figure 5.2
Gas-plant investment with no wind premium.

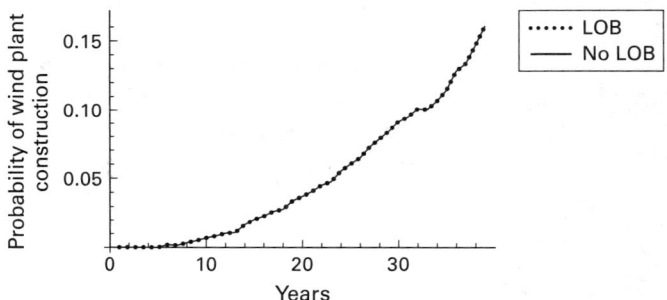

Figure 5.3
Investment probability wind with no wind premium.

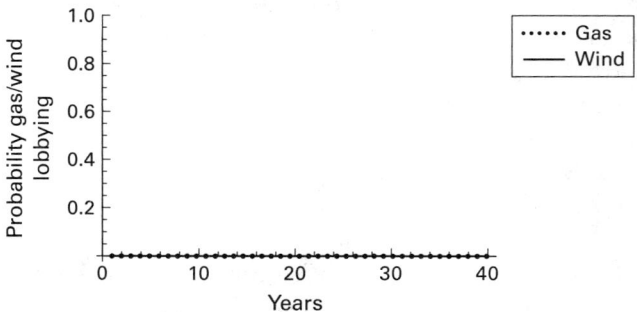

Figure 5.4
Lobbying probability with no wind premium.

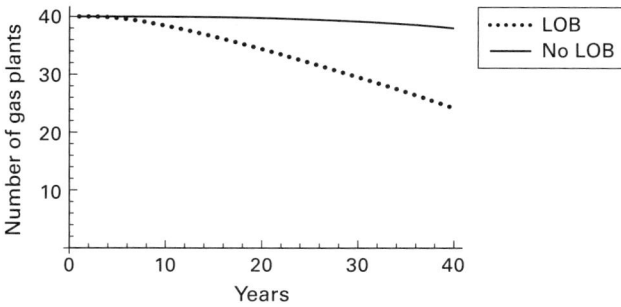

Figure 5.5
Gas-plant investment with wind premium.

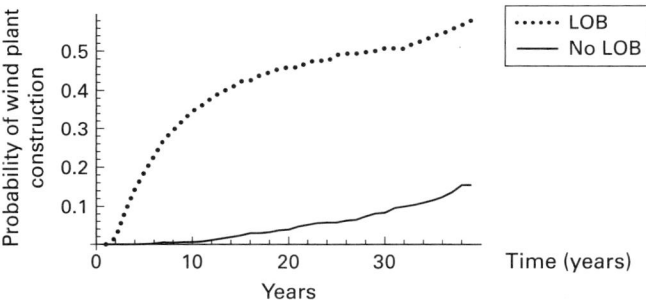

Figure 5.6
Investment probability wind with wind premium.

over the whole time horizon. The potential reduction in the permit price does not seem sufficient to justify the expense of lobbying.

In figures 5.5–5.8 we report the results for the case when both types of lobbying are allowed, and the wind premium is equal to €20. As can be seen in figure 5.5, wind power becomes the replacement choice after a few years, and by year 40 more than one fourth of the energy production is carried out by wind energy. In figure 5.6 we report the investment probability for wind. It quickly rises and reaches more than 50 percent by the end of the time horizon. The probability of wind investment with the opportunity to lobby is significantly higher than the without lobbying, as indicated by the area between the dotted and the solid lines in figure 5.5. Wind lobbying becomes a profitable choice, and lobbying activity rises quickly, reaching 70 percent toward the end of the time horizon. Gas lobbying is not carried out at all in this scenario. In figure 5.8 the CO_2 price is depicted. The price is the same with

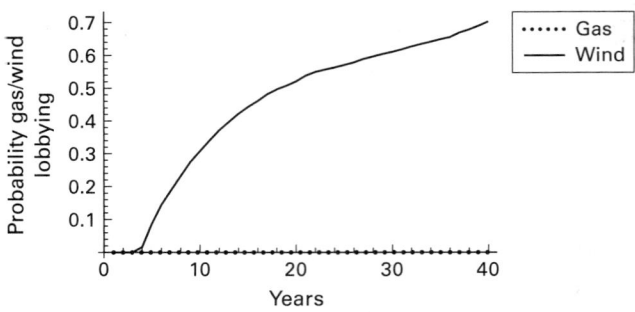

Figure 5.7
Lobbying probability with wind premium.

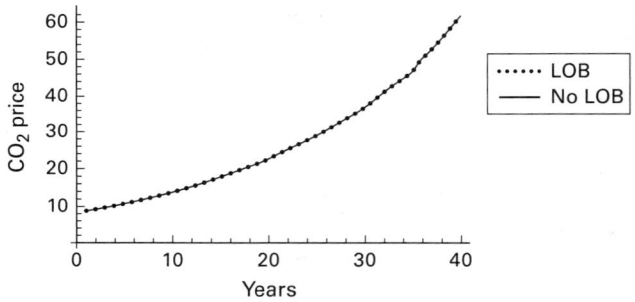

Figure 5.8
CO_2 price with wind premium.

or without lobbying, since no gas lobbying that would affect the prices of permits is carried out. The increase over time stems from the price process by which the prices of permits are driven.

Figures 5.9–5.11 show the level of incentives that is necessary to stimulate gas lobbying. The cost of lobbying is 10 percent of the cost of investment in gas power, the permit-price reduction achieved by successful gas lobbying is 75 percent, and the wind-power premium is €10. In this scenario, investment in gas power remains the dominant choice, since only at the end of the time horizon does investment in wind power occur. More important, lobbying for gas power dominates lobbying for wind power over the whole time period (figure 5.10), which leads to a significant drop in the CO_2 price over time (figure 5.11).

Finally, we present two scenarios to make a comparison between the European Energy Markets with its feed-in tariffs and the US system

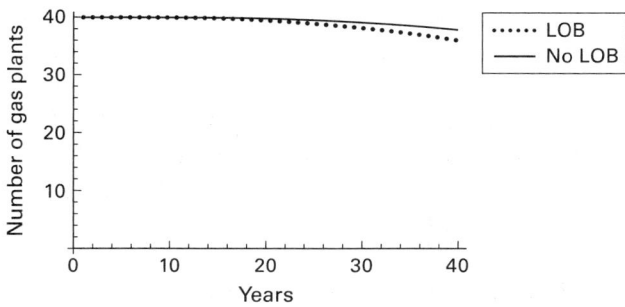

Figure 5.9
Gas-plant investment with gas lobbying.

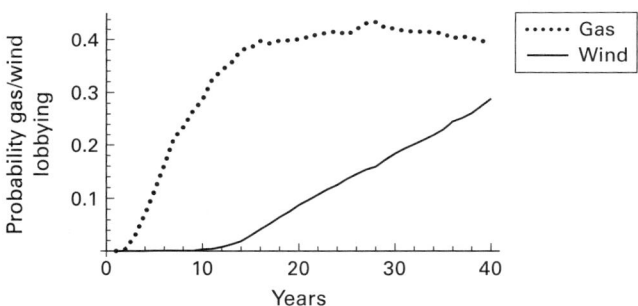

Figure 5.10
Lobbying probability with gas lobbying.

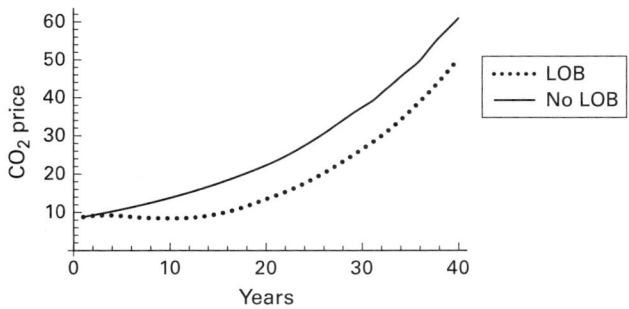

Figure 5.11
CO_2 price with gas lobbying.

with no CO_2 markets or substantial renewable subsidies. The main renewable energy support mechanism in the US are production tax credits, which have not been considered as reliable, mainly because of an uncertain renewal policy (Lewis and Wiser 2007; Barradale 2010). In both scenarios, lobbying for gas power is able to reduce the permit price by 100 percent. This implies that lobbying for gas power is able to completely offset the cost of CO_2 emissions; put differently, it can completely stop the establishment of an effective permit market. The first scenario, with lobbying for gas power 100 percent effective and with no renewable subsidy, is depicted in figures 5.12–5.14. In the second scenario, we also allow for a wind-power premium of €20 in order to show the difference such a mechanism would cause. The results of this scenario are shown in figures 5.15–5.17.

Without a renewable subsidy and with lobbying for gas power 100 percent effective, nearly no renewable energy infrastructure is installed

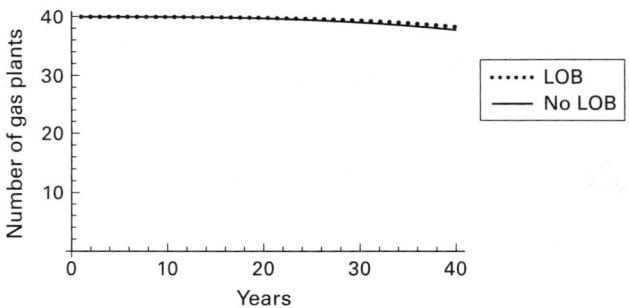

Figure 5.12
US gas-plant investment without subsidy.

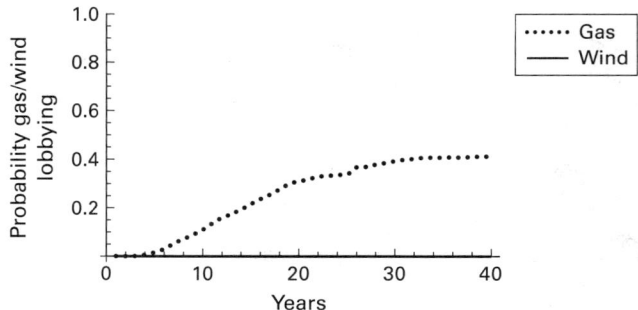

Figure 5.13
Lobbying probability: US scenario without subsidy.

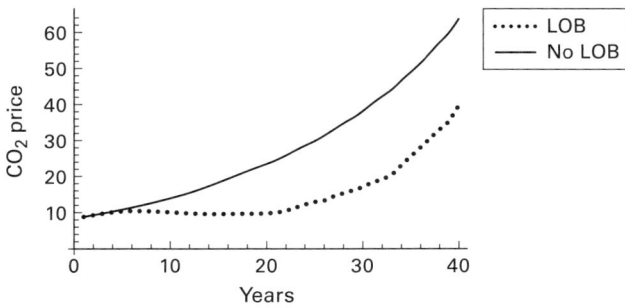

Figure 5.14
CO_2 price: US scenario without subsidy.

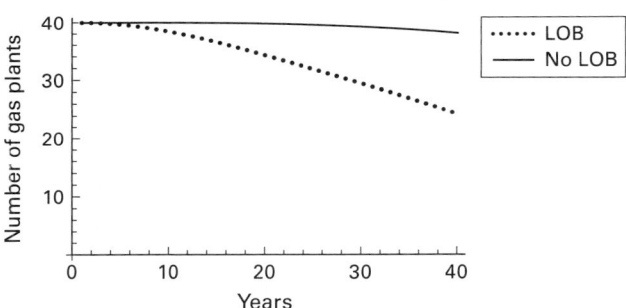

Figure 5.15
Gas-plant investment: US scenario with subsidy.

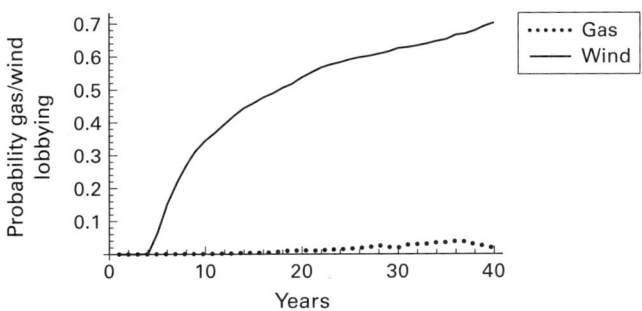

Figure 5.16
Lobbying probability: US scenario with subsidy.

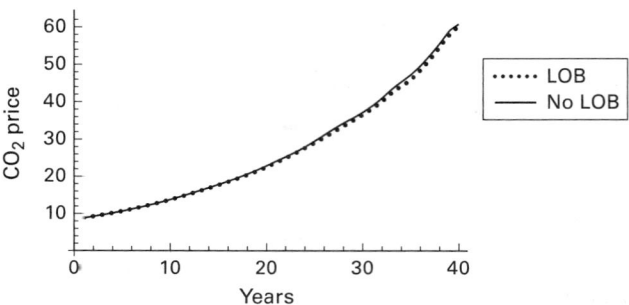

Figure 5.17
CO_2 price: US scenario with subsidy.

after 40 years (figure 5.12). Furthermore, after the CO_2 price reaches approximately €10 lobbying for gas power quickly begins to develop, reaching a probability of 0.4 by the end of the time period (figure 5.13). This heavily influences the CO_2 price, which remains around €10 until year 30, when it begins to increase again (figure 5.14).

When a renewable subsidy of €20 is introduced, even a 100 percent permit-price reduction is not sufficient to render gas power more profitable than wind energy. At the end of the time period, about one fourth of the energy production stems from renewable energy. (See figure 5.24.) This is very similar to the results obtained in the "wind premium" scenario, where lobbying for gas power was only 20 percent effectives. (See figure 5.5.) There is some lobbying for gas, though not as much as there is for wind power (figure 5.16). Therefore, the CO_2 price with lobbying differs minimally from that without lobbying (figure 5.17).

Sensitivity checks
In this subsection we report a range of sensitivity checks with respect to the interest rate, the volatility, the price of gas, the trend of the price process, the wind-energy premium, and the potential permit-price reduction due to the lobbying effort. As a baseline for comparison we choose the following values: the cost of lobbying is 20 percent of the gas-plant investment cost, the permit-price reduction is 20 percent, and the wind-power premium is €20. In the remaining figures, we lower the value we test for sensitivity to the left and increase it to the right.

The effect of changing the interest rates on gas investment from 1 percent to 5 percent to 9 percent can be seen in figures 5.18 and 5.19 (and in figure 5.5), the effect on the lobbying effort in figures 5.20 and

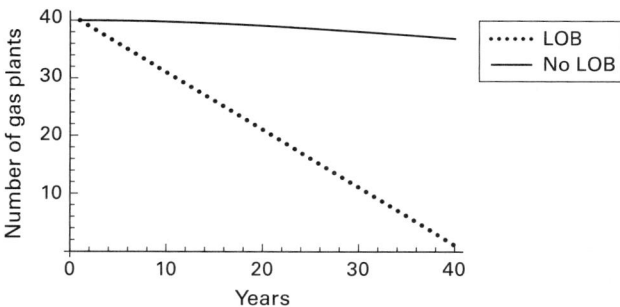

Figure 5.18
Interest rate 1 percent.

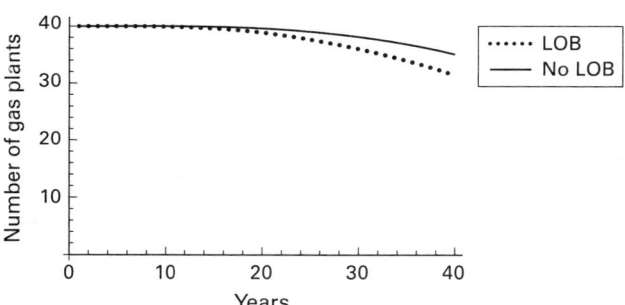

Figure 5.19
Interest rate 9 percent.

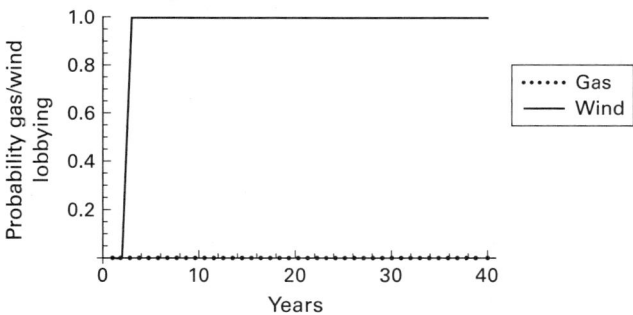

Figure 5.20
Interest rate 1 percent.

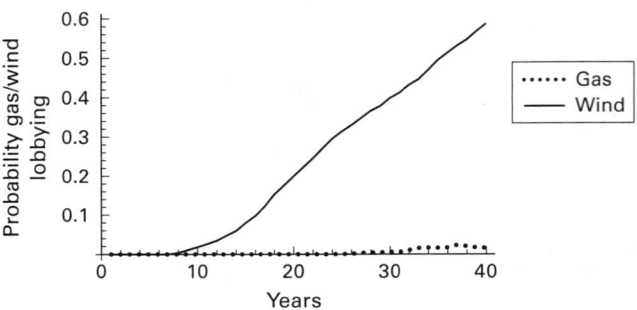

Figure 5.21
Interest rate 5 percent.

5.21 (and in figure 5.7). A 1 percent interest rate leads to a complete replacement of the original 40 gas power plants by the end of the time horizon with lobbying activity. This is due to the fact that a low interest rate favors wind power, since this type of power generation becomes more profitable relative the gas power the higher the permit price. Since the price process follows a geometric Brownian motion with a positive trend, the effect increases the further we move along in time. Consequently, at a 9 percent interest rate there are only few wind power plants installed by the end of the time horizon.

In lobbying behavior, a pattern similar to previous results can be observed. When wind power is very profitable and is installed early, it pays to invest in lobbying efforts for wind energy (interest rate 1 percent, figure 5.20). When gas power is more profitable, lobbying for wind power still dominates gas lobbying (interest rate 9 percent, figure 5.21), but wind power will remain as the only investment choice.

In figures 5.22 and 5.23 (and figure 5.5) we show the results for differing wind premium values on the number of gas power plants installed. The difference caused by decreasing/increasing the wind premiums is substantial with respect to the amount of installed renewable capacity by the end of the time horizon. Whereas a premium of €10 leads to an installed capacity of less than 10 percent, a €20 premium already leads to one fourth of installed renewable capacity, and a €30 premium leads to 100 percent of renewable capacity in year 40.

Figures 5.24 and 5.25 (and also figure 5.7) show the lobbying activity in these scenarios. Whereas a premium of €10 leads only to a wind lobbying probability of 30 percent by the end of the time horizon, a premium of €30 increases this probability to one over nearly the entire

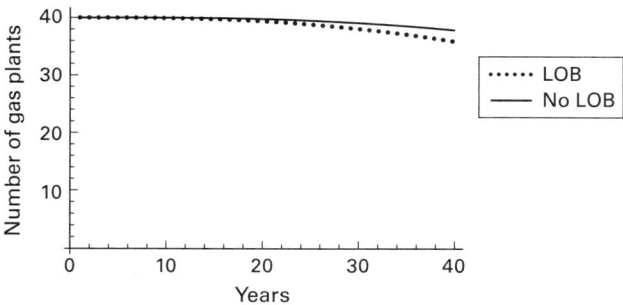

Figure 5.22
Gas-plant investment with €10 wind premium.

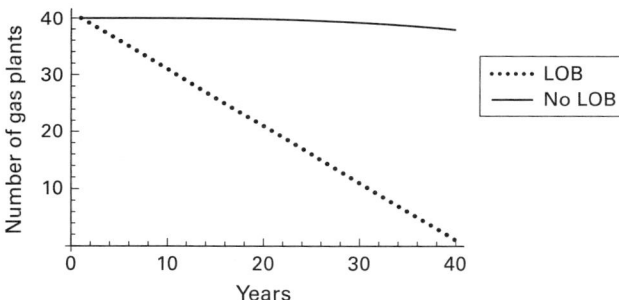

Figure 5.23
Gas-plant investment with €30 Wind Premium.

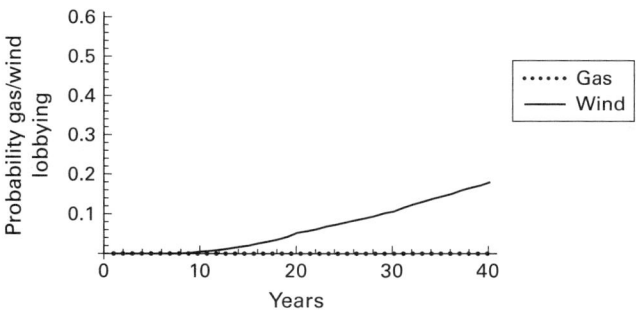

Figure 5.24
Lobbying probability with €10 wind premium.

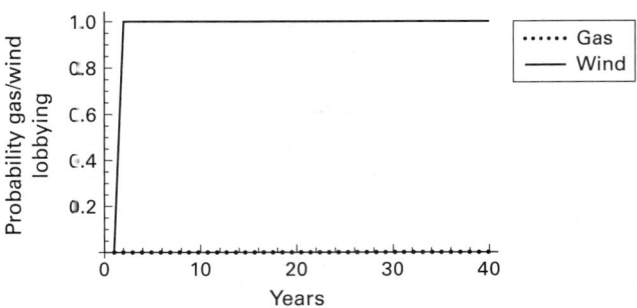

Figure 5.25
Lobbying probability with €30 wind premium.

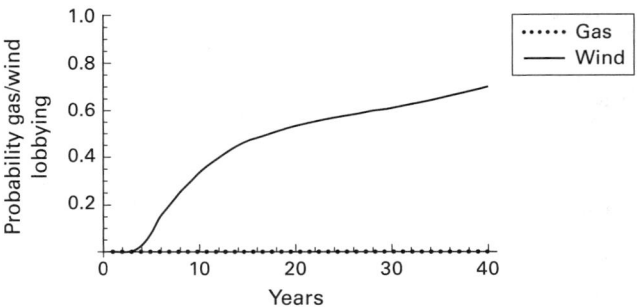

Figure 5.26
Lobbying probability (permit-price reduction 10 percent)

time horizon. Lobbying in order to increase the profitability of gas power plants becomes entirely unprofitable in all scenarios.

In figures 5.26 and 5.27 (and also figure 5.7) we report how lobbying affects the permit-price reduction. We report only the lobbying probability, since the investment probability changes only marginally with respect to the baseline scenario. The figures show that when an investor can receive a premium of €20 for wind energy or a 0.1–03 permit-price reduction, an investor rarely chooses to lobby for fossil energy. Lobbying for wind power behavior does not change substantially when varying the permit reduction rate.

In figures 5.28–5.31 (and also figures 5.5 and 5.7) we report the effect of changing the volatility on investment and lobbying behavior. The higher the volatility the more gas power plants are installed, even with a premium of €20 for wind energy. In the context of our model, a high

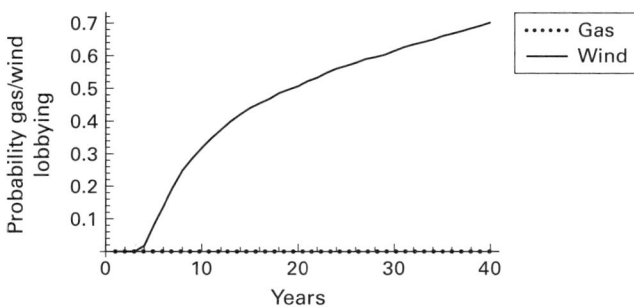

Figure 5.27
Lobbying probability (permit-price reduction 30 percent).

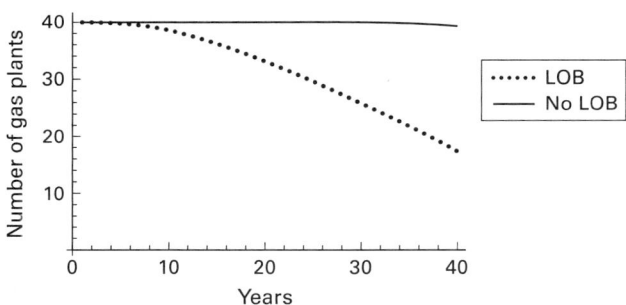

Figure 5.28
Gas power investment (volatility 10 percent).

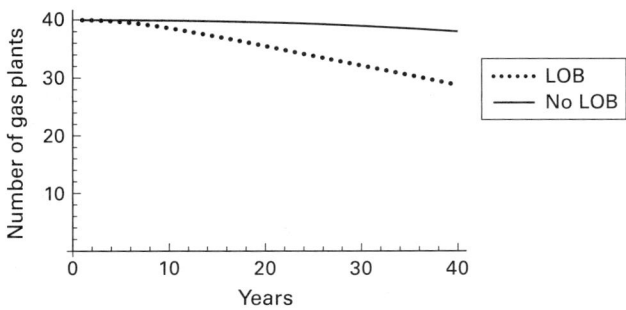

Figure 5.29
Gas power investment (volatility 30 percent).

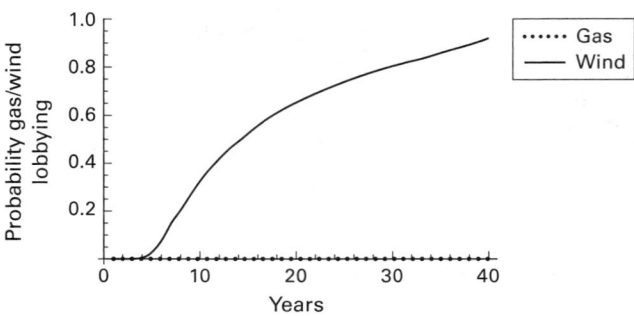

Figure 5.30
Lobbying probability (volatility 10 percent).

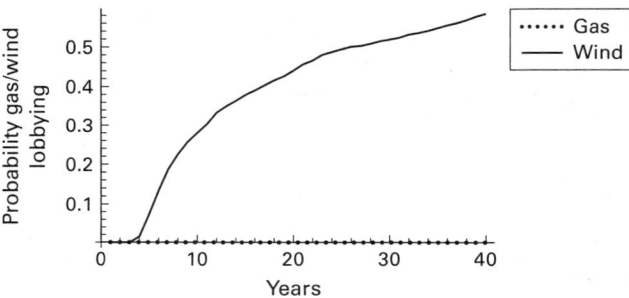

Figure 5.31
Lobbying probability (volatility 30 percent).

volatility implies a higher chance in later periods that the emission price can be very low, rendering investment in gas power more likely. Lobbying behavior (figures 5.30–5.31) is not much different than before.

In figures 5.32 and 5.33 (and also figure 5.5) we show how changing the gas price affects investment behavior. Figures 5.34 and 5.35 show the effect on lobbying behavior. Changing the gas price to €10 renders gas power highly profitable, and nearly no renewable energy is installed by the end of the time period, despite a wind premium of €20 (figure 5.32). A gas price of €50 has the opposite effect, by the end of the time horizon nearly all of the installed capacity comes from wind energy (figure 5.29). These results show that the price of gas affects the results of our model; however, changing the values to €10 or €50 is arguably a greater price change than is usually observed in the gas market, at least in the short to medium term.

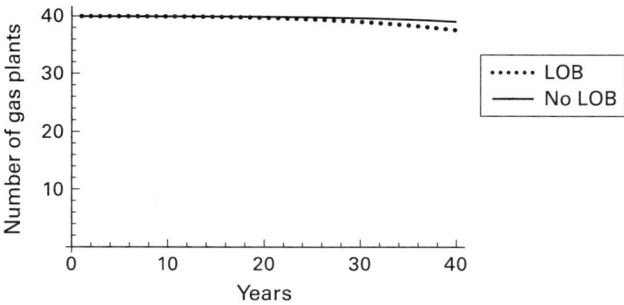

Figure 5.32
Gas power investment (gas price €10).

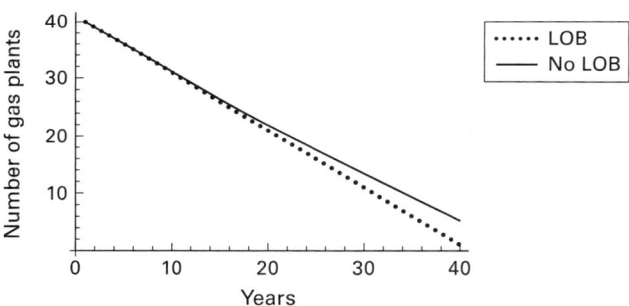

Figure 5.33
Gas power investment (gas price €50).

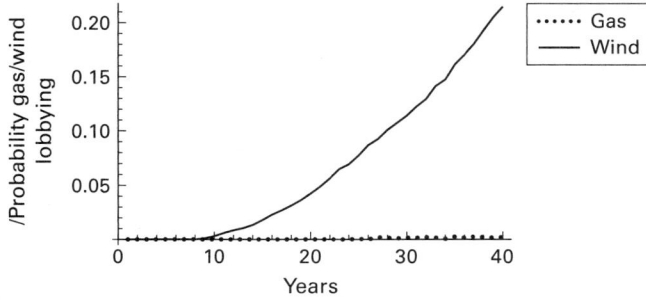

Figure 5.34
Lobbying probability (gas price €10).

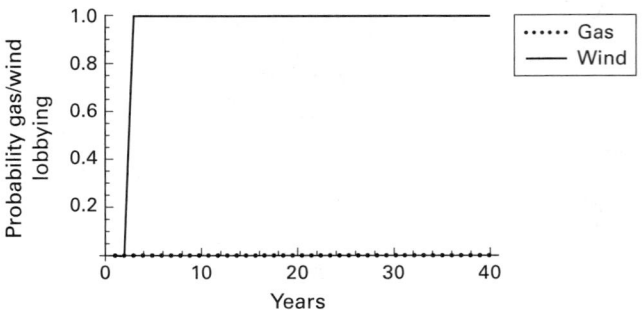

Figure 5.35
Lobbying probability (gas price €50).

Concerning lobbying behavior, a low gas price leads to a very low level of lobbying activity for both wind and gas. A high gas price renders lobbying for wind power very lucrative, since a large quantity of wind power is installed quickly and renders lobbying for wind power—already more lucrative than lobbying for gas power—even more profitable.

We also tested for a range of values for the price trend (1 percent, 5 percent, 10 percent) and for the lobbying cost (10 percent, 20 percent, 30 percent). None of the changes resulted in a significant deviation from previous results concerning investment and lobbying behavior.

Discussion

The results we have presented so far indicate that wind lobbying is the preferred choice, as CO_2 prices seem to be too low to warrant lobbying for permit-price reductions. However, the size of the wind-power premium has to be substantial in order to change the energy mix by the end of our time horizon. A premium of €20 (one third of the starting electricity price in euros) leads to 25 percent of installed wind energy capacity; a premium of €30 leads to 100 percent. Thus, at current permit prices, wind lobbying must achieve a substantial effect in order to change from a fossil to a renewable energy infrastructure. Because our model starts with 40 installed gas power plants, wind lobbying has to be significantly more profitable per power plant to change the energy infrastructure, as lobbying makes each power plant more profitable. We also reported the level of permit-price reduction and lobbying cost necessary in order to provide sufficient incentives for energy producers to lobby for a permit-price reduction. In a scenario with a wind-energy premium of €10, a lobbying cost of 10 percent relative to investment in gas power and a permit-price reduction of 75 percent lead to significant lobbying behavior for permit-price reductions. Whether such a level is reasonable for the effect of lobbying is certainly open to debate, but this gives an indication of the necessary effect that lobbying for fossil fuels would have to have in order to be profitable. Furthermore, we presented two scenarios in order to make a comparison between the US energy market with no CO_2 market and renewable subsidies and the European market. In order to emulate the US market within our model, we changed the permit reduction to 100 percent and removed the renewable energy lobbying option. In such a model, nearly no renewable energy will be installed by the end of the time horizon, gas lobbying becomes profitable when the CO_2 price reaches approximately €10, and the CO_2 price stays around this level over the remaining time period due to lobbying activity. We then allowed for renewable lobbying which yielded a €20 premium if successful, and found that even though gas lobbying took place with a certain probability and reduced the permit price toward the end of the time horizon, more than one fourth of the installed energy generating capacity was renewable. This implies that a significant renewable subsidy can overcome strong gas lobbying measures, at currently low CO_2 prices.

Concerning the sensitivity tests, we found that the interest rate, the wind premium, the volatility and gas price have the largest effect on our results. The interest rate is the most significant one as it affects the long-term investment prospects to a large degree. This implies that policy makers must be well informed about the interest rate in an investment environment if they want to judge the effects of changes in policy. Another factor that is mostly influenced by policy makers is the volatility rate of permit prices. Permit markets are still strongly dependent on political decisions, and therefore policy makers have a great level of influence on the volatility rate. For that reason, it is important to consider how policy changes affect volatility.

Even though we focus on the profitability of companies, leaving social welfare aside, it is worthwhile to analyze the rough magnitude of cost that the different scenarios imply. When only lobbying for gas power is possible, the social planner suffers from a reduced revenue since the CO_2 price can be depressed which leads to lower permit auction income. Allowing for lobbying for wind power renders gas lobbying unprofitable in most scenarios. Therefore, these cost no longer occur. However, the cost to society for the wind-power subsidy and the higher cost for installing the considerably more expensive energy type might be substantial. Assuming that there is no global carbon market, renewable capacity installed in one region of the world does not imply overall lower CO_2 emissions due to carbon leakage. Avoided damages due to a reduced externality level would then not factor in. One should also consider the potentially lower fuel cost as less fossil energy is consumed, which increases the incentive for fossil-fuel investment again. In conclusion, providing a wind-power premium may entail substantial economic cost for society as a whole. Policy makers will have to decide whether the benefits of renewable energy production, less CO_2 emissions at least in individual countries, a lower long-term energy price as production cost decrease, and jobs related to renewable energy production are worth the cost.

Conclusion

We have presented a modeling framework to answer the following questions:

• How does lobbying for wind or gas power affect the investment decision under permit-price uncertainty?
• When is the option to lobby used?
• Is the wind or gas lobbying option used?

We find that lobbying for permit-price reductions at the currently low CO_2 price level is only profitable if the potential reduction is very large (75–100 percent) and the wind-power premium is small (€10If the wind-power premium is €20, about one fourth of the installed energy stems from wind by the end of our 40 year time period, rising to 100 percent for a premium of €30. Furthermore, wind lobbying activity dominates any gas lobbying activity in such a scenario. The effect of wind lobbying is also substantial, in that the difference between a wind lobbying and no lobbying scenario in terms of investment probability into a wind power plant is up to 100 percent. We also considered a scenario emulating the situation in the US market with no CO_2 market and no effective renewable support tool. In this scenario we allowed for a 100 percent effectiveness of gas lobbying, implying that lobbying can completely stop the operation of a CO_2 market. We found that in such a scenario the amount of renewable energy is negligible by the end of the time period. Furthermore, lobbying for gas power still only becomes profitable when the CO_2 price reaches a level of around €10.

When introducing a wind-energy premium of €20 into such a system, a 25 percent renewable share in the energy mix can be reached. We performed a range of sensitivity checks with respect to the interest rate, volatility, trend of permit prices, gas price, the cost of lobbying, and effectiveness of lobbying. The interest rate plays a crucial role since a higher interest rate discounts the future more, and leads to a less favorable investment environment for wind power since this power type becomes more profitable over time relative to gas power. The volatility also affected the results in a significant fashion since a higher volatility may lead to low permit prices in later periods, rendering gas power more profitable.

Even though our model is a partial equilibrium analysis we gave an indication of the social welfare cost attached to the different scenarios. Under the assumption that a 5 percent permit-price trend adequately captures the value of the negative environmental externality caused by CO_2 emissions, allowing for either type of lobbying implies a welfare loss. The scenarios clearly showed that a 5 percent trend coupled with currently low permit prices is not sufficient in order to induce investment in renewable energy. With the introduction of a feed-in tariff like support mechanism for wind power, the amount of renewable energy increased substantially. This support, however, comes at a cost. In addition to the cost for the subsidy, wind power is the more expensive energy type leading to higher cost for society. Furthermore, costly wind lobbying activity becomes profitable implying further cost. Without a global emission market carbon leakage can become an issue. Over time, increasing high-cost renewable energy production in one part of the world may cause energy-intensive production to move elsewhere or even increase incentives to use fossil resources via reduced market prices due to less consumption from renewable-intensive countries.

The quantitative results of our model should be treated with care, since the 5 percent trend is the current best guess of what a future permit might like look like. The same argument applies to the lobbying cost. Therefore, it is the general direction and relative size our results indicate which are the main result of our study, not specific year for year values. We have abstracted from cartel and oligopoly issues which certainly play a role in the permit market. Also, a thorough general equilibrium welfare analysis of the implied social costs are beyond the scope of our study. Addressing these issues is left for future research.

Acknowledgments

We thank Ralf Martin, Thorsten Upmann, the participants in the 26th PhD Workshop in Climate Policy, the CESifo Venice Summer Institute Workshop on "Emission Trading Systems as a Climate policy Instrument: Evaluation and Prospects," the Cologne International Energy Summer in Energy in Environmental Economics, and the pERE Seminar at University of Illinois at Urbana-Champaign for valuable comments and suggestions. Funding from CESifo, the Institute of Energy Economics at the University of Cologne and the German Academic Exchange Service is gratefully acknowledged by David Schüller.

Notes

1. http://www.ewea.org/about-us/

2. See, for example, Insley 2003; Abadie and Chamorro 2006, 2008; Yang et al. 2008; Fuss et al. 2009, 2010; Abadie and Galarraga 2011. For a more elaborate overview, see Hieronymi and Schüller 2015.

3. We focus on the most common element of lobbying in the US and the EU: that it costs money. Thereby we abstract from difference in the two lobbying systems with respect to such dimensions as trust (Woll 2006; Greenwood 2011).

4. All euro values are converted into US dollars at an exchange rate of $1.35 per euro.

5. "The overnight cost therefore excludes escalation in equipment, labor, and commodity prices that could occur during the time a plant is under construction. It also excludes the financing charges, often referred to as interest during construction (IDC), incurred while the plant is being built." (Kaplan 2008, p. 696)

6. All data stemming from the EEX website (http://www.eex.com/en/), including the electricity, gas, and permit prices, were downloaded on January 22, 2014.

7. All euro values are converted into US dollar values at an exchange rate of $1.35 per euro.

8. EUA: €6,52; data from http://www.eex.com converted into US dollars.

9. http://www.bloomberg.com/news/2012-03-15/eu-parliament-calls-for-emissions-permt-set-aside-option-1-.html

10. http://www.opensecrets.org/lobby/

11. 400 MW × 1,000 (KW conversion) × US$1,068 (overnight cost) × 0.85 (capacity factor) = US$427 million.

12. http://www.ecoprog.com/en/publications/energy-industry/gas-power-plants .htm

13. http://www.eia.gov/electricity/annual/pdf/table5.1.pdf

14. http://www.eia.gov/electricity/annual/pdf/tables1.pdf

References

Abadie, L., and J. Chamorro. 2006. Monte Carlo valuation of natural gas investments. *Energy Economics* 18 (1): 10–22.

Abadie, L., and J. Chamorro. 2008. European CO_2 prices and carbon capture investments. *Energy Economics* 30 (6): 2992–3015.

Abadie, L., and I. Galarraga. 2011. The European Emission Trading Scheme: Implications for long-term investment valuation. *Climate Change Economics* 2 (2): 129–148.

Barradale, M. 2010. Impact of public policy uncertainty on renewable energy investment: Wind power and the production tax credit. *Energy Policy* 38 (12): 7698–7709.

Becker, G. 1983. A theory of competition among pressure groups for political influence. *Quarterly Journal of Economics* 98 (3): 371–400.

Becker, G. 1985. Public policies, pressure groups, and dead weight costs. *Journal of Public Economics* 28: 329–347.

Black, F., and M. Scholes. 1973. The pricing of options and corporate liabilities. *Journal of Political Economy* 81 (3): 637–654.

Boogert, A., and C. de Jong. 2011. Gas storage valuation using a multi-factor price process. *Journal of Energy Markets* 4 (4): 29–52.

Brandt, U., and G. Svendsen. 2004. Fighting windmills: The coalition of industrialists and environmentalists in the climate change issue. *International Environmental Agreement: Politics, Law and Economics* 4 (4): 327–337.

Conconi, P. 2003. Green lobbies and transboundary pollution in large open economies. *Journal of International Economics* 59 (2): 399–422.

Dixit, A., and R. Pindyck. 1994. *Investment Under Uncertainty*. Princeton University Press.

EEX (European Energy Exchange). 2014. http://www.eex.com/en/.

Fezzi, C., and D. Bunn. 2009. Structural interaction of European carbon trading and electricity prices. *Journal of Energy Markets* 2 (4): 53–69.

Fredriksson, P., E. Neumayer, S. Gates, and G. Per. 2005. Environmentalism, democracy and pollution control. *Journal of Environmental Economics and Management* 49 (2): 343–365.

Fuss, S., J. Szolgayova, M. Obersteiner, and M. Gusti. 2008. Investment under market and climate policy uncertainty. *Applied Energy* 85 (8): 708–721.

Fuss, S., D. Johansson, J. Szolgayova, and M. Obersteiner. 2009. Impact of climate policy uncertainty on the adoption of electricity generating technologies. *Energy Policy* 37 (2): 733–743.

Fuss, S., J. Szolgayova, A. Golub, and M. Obersteiner. 2010. Options on low-cost abatement and investment in the energy sector: New perspectives on REDD. *Environment and Development Economics* 16 (4): 1–19.

Greenwood, Justin. 2011. *Interest Representation in the European Union*, third edition. Palgrave Macmillan.

Grossman, G. M., and E. Helpman. 1994. Protection for sale. *American Economic Review* 84: 833–850.

Gulbrandsen, L., and S. Andresen. 2004. NGO influence in the implementation of the Kyoto Protocol: Compliance, flexibility mechanisms, and sinks. *Global Environmental Politics* 4 (2): 75–99.

Gullberg, A. 2008. Lobbying friends and foes in climate policy: The case of business and environmental interest groups in the European Union. *Energy Policy* 36 (8): 2964–2972.

Habla, W., and R. Winkler. 2013. Political influence on non-cooperative international climate policy. *Journal of Environmental Economics and Management* 66: 219–234.

Hanley, N., and I. Mackenzie. 2010. The effects of rent seeking over tradable pollution permits. *B.E. Journal of Economic Analysis & Policy* 10 (1): 1–26.

Helm, D. 2010. Government failure, rent-seeking, and capture: The design of climate change policy. *Oxford Review of Economic Policy* 26 (2): 182–196.

Hieronymi, P., and D. Schüller. 2015. The Clean-Development Mechanism, stochastic permit prices and energy investments. *Energy Economics* 47: 25–36.

IEA. 2010. *Projected Costs of Generating Electricity*, 2010 edition.

Insley, M. C. 2003. On the option to invest in pollution control under a regime of tradable emissions allowances. *Canadian Journal of Economics. Revue Canadienne d'Economique* 36 (4): 860–883.

Kaplan, S. 2008. *Power Plants: Characteristics and Costs*. Congressional Research Service.

Keppler, J., and M. Cruciani. 2010. Rents in the European power sector due to carbon trading. *Energy Policy* 38 (8): 4280–4290.

Kettunen, J., D. Bunn, and W. Blyth. 2011. Investment propensities under carbon policy uncertainty. *Energy Journal* 32 (1): 77–118.

Lai, Y. 2006. The optimal distribution of pollution rights in the presence of political distortions. *Environmental and Resource Economics* 36 (3): 367–388.

Lai, Y. 2008. Auctions or grandfathering: The political economy of tradable emission permits. *Public Choice* 136 (1–2): 181–200.

Lewis, J., and R. Wiser. 2007. Fostering a renewable energy technology industry: An international comparison of wind industry policy support mechanisms. *Energy Policy* 35 (3): 1844–1857.

Markussen, P., and G. Svendsen. 2005. Industry lobbying and the political economy of GHG trade in the European Union. *Energy Policy* 33 (2): 245–255.

Merton, R. 1973. Theory of rational option pricing. *Bell Journal of Economics and Management Science* 4 (1): 141–183.

Myers, S. C. 1977. The determinants of corporate borrowing. *Journal of Financial Economics* 5: 147–175.

Skodvin, T., A. Gullberg, and S. Aakre. 2010. Target-group influence and political feasibility: The case of climate policy design in Europe. *Journal of European Public Policy* 17 (6): 854–873.

Svendsen, G. 1999. U.S. interest groups prefer emission trading: A new perspective. *Public Choice* 101 (1–2): 109–128.

Trigeorgis, L. 1996. *Real Options: Managerial Flexibility and Strategy in Resource Allocations.* MIT Press.

Wettestad, J. 2009. EU energy-intensive industries and emission trading: Losers becoming winners? *Environmental Policy And Governance* 320: 309–320.

Woll, Cornelia. 2006. Lobbying in the European Union: From sui generis to a comparative perspective. *Journal of European Public Policy* 13 (3): 456–469.

Yang, M., W. Blyth, R. Bradley, D. Bunn, C. Clarke, and T. Wilson. 2008. Evaluating the power investment options with uncertainty in climate policy. *Energy Economics* 30 (4): 1933–1950.

III Overlapping Policy Instruments

6 Combining International Cap-and-Trade with National Carbon Taxes

Peter Heindl, Peter J. Wood, and Frank Jotzo

This chapter examines linkages of price and quantity instruments for reductions in emissions. Regulation by prices is introduced as an additional tax on top of regulation by quantities. The model builds on a more general analysis of price-quantity linking for public-good provision in Wood et al. 2013.

We assume that two countries are covered by a joint cap-and-trade scheme and aim to comply with their respective targets for emissions levels, which are below expected business-as-usual emissions. In addition to the cap-and-trade scheme a tax is levied in country 1, on top of the cap-and-trade price. Allowances (or permits) are fully tradable, so compliance with the emissions targets occurs in aggregate: if one country reduces its emissions below the target, the other country can stay above its target, and the difference is accounted for by permits traded between the two countries. We take uncertainty in the cost of public good production into account by applying log-normally distributed error terms to the countries' cost functions.

Situations as modeled in this chapter occur in practice. A prominent example is the EU Emissions Trading Scheme. The EU ETS is a joint cap-and-trade scheme of 27 EU member states, Norway, and Liechtenstein. Exogenous abatement targets are implicitly given by the countries' commitments to reduce emissions—for example, by the EU burden-sharing agreement (EU 2002/358/EC 2002). In addition to the joint trading scheme, some countries introduced additional policies. Examples are the UK "price floor," which works as an additional charge on emissions for the UK power sector, to bring the effective carbon price to a predetermined minimum level (Toke 2011); carbon taxes that have existed in several European countries in parallel to the EU ETS (OECD 2013); as well as the many non-pricing policies that exist in different countries, such as the German renewable feed-in tariff system (IEA 2013). We model the partial equilibrium effect of such policies on prices, costs, and abatement in a joint trading scheme.

The existence of a "single price" on some externality is a condition for efficiency in many situations. If firms minimize costs, and if there are no market distortions, marginal costs will be equated among all sources of emissions given the single price, which is a least-cost solution (Baumol and Oates 1971, 1988; Montgomery 1972). Thus it is to be expected that the overall cost of reducing emissions is higher if the marginal cost of abatement effort differs between countries. This has been shown in general equilibrium modeling exercises such as those conducted by Böhringer, Löschel, et al. (2009), who suggested that the overlapping instruments and multiple targets of the EU approach will lead to excess cost against a first best situation; by Böhringer, Rutherford, et al. (2009), who showed that additional policies (i.e., multiple carbon prices) will likely increase the overall cost of greenhouse-gas mitigation in the EU considerably; and Boeters and Koornneef (2011), who examined the likely cost of the EU renewable energy target, which is a separate element of the EU 2020 climate strategy, in parallel to the emissions-reduction target.

It is important to note that the single price that would result from an emissions trading scheme need not necessarily equal the "Pigouvian" price that equates the marginal cost of abatement with its marginal benefit. In practice the social costs of an externality such as greenhouse-gas emissions are difficult to assess (Baumol 1972). It may be that additional policies beyond the EU ETS, such as the UK carbon price floor or the German renewable feed-in tariff scheme, reflect different views about acceptable or appropriate levels of externalities or costs of emissions abatement. In this chapter, we set aside questions of the (social) optimality of emissions abatement targets, and instead focus on the distribution of costs of achieving some policy target. Our model is stochastic, allowing assessment of the probability distribution of different outcomes for costs, and thus an assessment of cost risks alongside expected values. Our modeling omits general equilibrium effects which can be of importance for assessments of actual (social) cost under different revenue recycling regimes or given pre-existing taxes (Bovenberg and Goulder 1996; Bovenberg and van der Ploeg 1994; Bovenberg 1999). Different revenue recycling regimes can also have major implications for first-best instrument choice (Pezzey and Jotzo 2012).

We show that an additional tax levied in one country will increase the country's cost of producing the public good, decrease the others country's cost, and increase overall costs. This is because of differences in marginal abatement costs. The allowance price in the joint cap-and-trade scheme will decrease as result of the tax relative to a situation without the extra tax. The relative size of countries and the magnitude and correlation of uncertainty are important determinants for the actual effect of the tax on the allowance price and total costs. Additional abatement generated by the tax will only occur in cases when all abatement is generated by the tax and exceeds the joint quantity target of public good production of both countries. This represents a corner solution in which the cap-and-trade allowance price equals zero. Expected costs can be far higher in this situation than in the case of pure cap-and-trade without the tax.

6.1 The model

The model considers two countries (or, alternatively, regions, sectors, or firms) which are jointly covered under quantity regulation by cap-and-trade. Each country has an emissions target, and needs to reduce its emissions below some business-as-usual level. This can be described as each country having been allocated the production of a quantity of a public good (e.g., emissions reductions), denoted by Q_i. The production of the good incurs costs, which are ex-ante uncertain. Country 1 or country 2 produces abatement $q_i \geq 0$ at costs c_i, so that

$$c_i = c_i(q_i, \theta_i),$$

where the random variable θ_i influences the costs for country i.[1]

Cost functions are continuous and twice differentiable. Partial derivatives of country i's cost function with respect to q satisfy

$$c'_q > 0$$

and

$$c''_q > 0.$$

Pure cap-and-trade
When both countries produce the amount of $Q_1 + Q_2$ under a joint cap-and-trade scheme, and assuming that there are no distortions in the market, marginal costs will be equated.[2] Allowances issued under

cap-and-trade are fully tradable and it is assumed that there is full compliance with the quantity targets. The joint amount of the produced good equals the required quantity, so that $q_1 + q_2 = Q_1 + Q_2$. If a is the amount of allowances sold from country 1 to country 2 (a can be positive or negative), the equilibrium market price for allowances p^* satisfies

$$p^* = c_1'(q_1 + a, \theta_1) = c_2'(q_2 - a, \theta_2). \tag{1}$$

Equation 1 characterizes the cost-minimizing allocation of abatement between the countries under cap-and-trade (in the absence of transaction costs, market power and pre-existing distortions). The allowance price equals the marginal valuation of allowances by the countries. The amount of traded allowances is chosen so that abatement costs are minimized in each of the countries. A country will sell allowances as long as marginal revenues from selling allowances are larger than marginal costs of domestic public good production. A country will purchase allowances as long as allowances are sold on the market at cost below the marginal cost of producing the public good domestically. Thus, for a given vector of public good production ($Q_1 + Q_2$), there is a cost-minimizing price p^* (Baumol and Oates 1971). With a "market in emission licenses," the market will reach the cost-minimizing equilibrium (Montgomery 1972).

In figure 6.1, the equilibrium allowance price under cap-and-trade is given at point A where marginal costs of both countries are equated.

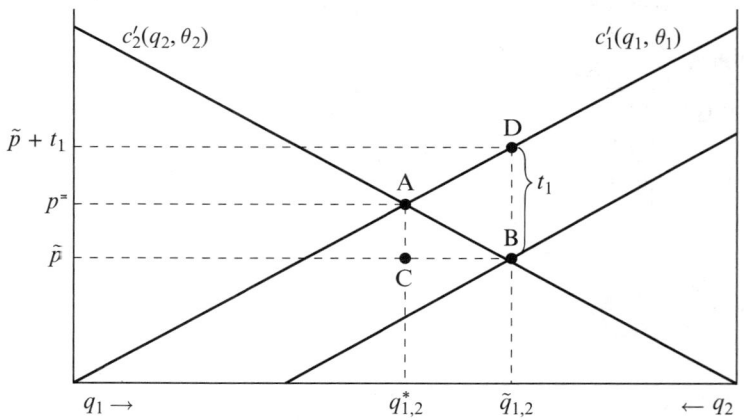

Figure 6.1
Marginal cost functions $c_i'(q_i, \theta_i)$ and produced quantities of both countries q_i (x axis) with and without the tax introduced by country 1.

Countries produce quantities $q_{1,2}^*$ at price p^*. In this case, no trading of allowances occurs.

Because marginal costs are equated between the countries, there are no options to further decrease the costs of producing the public good. If both countries choose the cost-minimizing quantities of the public good production q_i^* so that it satisfies equation 1 and subject to the quantity constraint $q_1 + q_2 = Q_1 + Q_2$, costs for each country are

$$C_1(\theta_1, \theta_2) = c_1(q_1^*, \theta_1) - p^* a \tag{2}$$

and

$$C_2(\theta_1, \theta_2) = c_2(q_2^*, \theta_2) + p^* a. \tag{3}$$

The case of "pure" cap-and-trade as described in this section will serve as reference case in the model in the case of cap-and-trade with an additional tax in country 1. Since a fully competitive cap-and-trade scheme with zero transaction costs is assumed as reference case, it is important to note that results could change considerably if these assumptions are relaxed. Market power of one of the parties engaged in trading will (in most cases) lead to a different allocation and distribution of cost than in the competitive case (Hahn 1984; Sinn and Schmoltzi 1981). Transaction cost, e.g., for allowance trading, will change the equilibrium allocation under cap-and-trade (Stavins 1995).

Additional unilateral tax by one country under cap-and-trade

We now turn to the case where country 1 implements a policy to produce a larger amount of the public good domestically relative to the existing cap-and-trade scheme. To do so, an additional unit-tax $t_1 > 0$ on top of the allowance price under cap-and-trade is levied in country 1.[3] The tax results in production of the same tradable good, namely emissions reductions in country 1, whether country 1 initially was a net exporter or a net importer of the good under the cap-and-trade regime. The price under cap-and-trade and the additional tax must satisfy

$$\tilde{p} = c_1'(\tilde{q}_1 + a, \theta_1) - t_1 = c_2'(\tilde{q}_2 - a, \theta_2). \tag{4}$$

The tilde denotes the cap-and-trade price and amounts of the public good provided if there is a tax, as distinct from the case without the additional tax above. Again both countries choose the optimal amounts of public good $\tilde{q}_i$ so that it satisfies equation 4 and subject to the quantity constraint $\tilde{q}_1 + \tilde{q}_2 = Q_1 + Q_2$ at tax rate t_1.

The amount of public good produced by country 1 consists of two parts: one attributable to the tax and the other attributable to the permit price under cap-and-trade. If the allowance price under cap-and-trade was equal to zero, country 1 would still produce some amount of the public good by virtue of the tax. The additional quantity produced by country 1 in response to the tax is equal to distance CB in figure 6.1 (in which the allowance price is positive).

The unilateral introduction of the tax by country 1 changes marginal conditions and, other things equal, leads to increased public good production in country 1, so that $\tilde{q}_1 > q_1^*$. Increased production by country 1 lowers the need for production by country 2 given the quantity constraint $Q_1 + Q_2$, so that $\tilde{q}_2 < q_2^*$. As a consequence, the allowance price under cap-and-trade decreases.

If $\tilde{q}_1 \geq Q_1 + Q_2$, there is no need for further public good production (emissions reductions) by country 2. This is a corner solution in which all of the good is produced in country 1. Country 1 values the public good by its marginal costs $c_1'(\tilde{q}_1, \theta_1) = p + t$. It follows from (4) that for $\tilde{q}_1 \geq Q_1 + Q_2$, the allowance price is zero and country 2 acquires Q_2 from country 1 to fulfill its target. Costs in the corner solution are

$$C_1(\theta_1, \vartheta_2) = c_1(\tilde{q}_1, \theta_1) \tag{5}$$

and

$$C_2(\theta_1, \theta_2) = 0. \tag{6}$$

Alternatively, if $\tilde{q}_1 < Q_1 + Q_2$, there is need for further abatement in the joint cap-and-trade scheme. In this case, countries produce the amounts of $\tilde{q}_i$ of the public good which satisfies equation 4. From equation 4 it follows that a positive valuation is given to the tradable public good either by country 1 (given the tax) or by country 2, so that $p^* > \tilde{p} > 0$. Since countries are assumed to comply with their quantity targets Q_1, Q_2, the number of traded permits in the interior solution is $a = Q_2 - \tilde{q}_2$.[4] Costs are

$$C_1(\theta_1, \theta_2) = c_1(\tilde{q}_1, \theta_1) - \tilde{p}a \tag{7}$$

and

$$C_2(\theta_1, \theta_2) = c_2(\tilde{q}_2, \theta_2) + \tilde{p}a. \tag{8}$$

Whenever $t_1 > 0$ with $\tilde{p} < p^*$, $\tilde{q}_1 > q_1^*$, and $\tilde{q}_2 < q_2^*$, overall cost of public good production exceed costs in the case of pure cap-and-trade. This is because

$$\int_{q_1^*}^{\tilde{q}_1} c_1' - \int_{\tilde{q}_2}^{q_2^*} c_2' > 0.$$

(In figure 6.1, excess costs are represented by the area BAD.) In other words, the cost increases in country 1 cannot be fully offset by cost decreases in country 2.

Because of the binding quantity constraint $Q_1 + Q_2$, no additional amount of the public good (exceeding the amount that would have been produced under pure cap-and-trade) will be produced in the interior solution. Additional amounts of the public good will only be produced in the corner solution where $\tilde{q}_1 > Q_1 + Q_2$. In this case country 1 produces all of the public good and the produced quantity can exceed the cap-and-trade quantity constraint $Q_1 + Q_2$.

Figure 6.1 illustrates that the introduction of the tax in country 1 shifts public good production, so that country 1 produces more of the good relative to pure cap-and-trade ($q_{1,2}^*$ to $\tilde{q}_{1,2}$). The introduction of the tax has two effects. Firstly, the effective price and the produced quantity in country 1 increase from A to D in figure 6.1. Secondly, the effective price and the quantity produced in country 2 decrease from A to B. This drives a wedge between marginal abatement costs and causes excess costs equal to the area BAD. The two effects are not necessarily symmetric as in figure 6.1. They depend on abatement cost parameters α_i of the countries relative to each other, targets Q_i, and uncertainty inherent in abatement cost functions. Below we examine the two effects by taking expectations over log-normally distributed "shocks" to abatement cost functions.

Under pure cap-and-trade, the price for allowances (y axis) is p^* and equals marginal abatement costs of both countries (A). Countries produce quantities q_i^*. If country 1 unilaterally levies t_1 (distance BD) on top of the allowance price, the allowance price will decrease to $\tilde{p}$. Country 1 will produce $\tilde{q}_1 > q_1^*$ and country 2 will produce $\tilde{q}_2 < q_2^*$. The tax-induced reallocation of public good production is equal to the distance CB and $\tilde{q}_1 + \tilde{q}_2$ is equal to $Q_1 + Q_2$. Overall costs will increase if country 1 introduces t_1 relative to pure cap-and-trade. Excess costs are represented by the area BAD. The overall quantity target $Q_1 + Q_2$ is equal to the width of the x axis and equal to $q_1^* + q_2^*$ or $\tilde{q}_1 + \tilde{q}_2$ respectively. The example shown in this figure is an interior solution. The marginal cost function of country 1 is shifted downward since functions are plotted in the space of cap-and-trade prices and quantities and country 1 would produce some of the good even if the cap-and-trade price (y axis) was zero.

Model parameterization

To gain empirical insights into the issue under uncertainty, cost functions are parameterized. We use the same parameterization for cost functions as in Wood et al. 2013. Abatement cost functions for countries 1 and 2 are given by

$$c_i\left(q_i,\theta_i\right)=\tfrac{1}{2}\theta_i\alpha_iq_i^2,$$

where α_i determines the slope of the marginal cost curves and is constant, θ_i is a random variable, and q_i is the amount of the produced abatement.

As in Wood et al. 2013, it is assumed that the random variable θ_i is log-normally distributed and has a mean value of 1. Marginal costs are given by

$$c_i'\left(q_i,\theta_i\right)=\theta_i\alpha_iq_i.$$

Assuming linear marginal costs keeps the analysis more tractable and has been standard procedure in the relevant literature (e.g., Hagem and Westskog 1998). Relative to abatement cost curves derived from empirical studies, linear marginal abatement costs will likely underestimate the increase in abatement costs as marginal costs increase, however this has no bearing on our qualitative results.[5]

In the interior solution, the number of allowances sold from country 1 to country 2, a, will be the choice for which

$$c_1'\left(Q_1+a,\theta_1\right)+t=c_2'\left(Q_2-a,\theta_2\right);$$

and for the corner solution, $a=Q_2$. The minimum tax rate for which there is a corner solution is the tax rate for which the number of allowances (in the interior solution) is equal to Q_2. It follows that with the parameterization above, the number of allowances is given by the following equation:

$$a=\begin{cases}\dfrac{-\alpha_1\theta_1Q_1+\alpha_2\theta_2Q_2+t}{\alpha_1\theta_1+\alpha_2\theta_2}, & t<\alpha_1\theta_1(Q_1+Q_2)\\[2mm] Q_2 & \text{otherwise.}\end{cases}$$

The allowance price is equal to the marginal cost for country 2 to produce the amount Q_2-a of abatement, and is given by the following equation:

$$p = \begin{cases} -\dfrac{\alpha_2\theta_2\,(\alpha_1\theta_1 Q_1 - \alpha_1\theta_1 Q_2 + t)}{\alpha_1\theta_1 + \alpha_2\theta_2}, & t < (Q_1 + Q_2)\alpha_1\theta_1 \\ 0 & \text{otherwise.} \end{cases}$$

The total costs for each country are then calculated by substituting a and p into the equations

$$C_1 = c_1(Q_1 + a, \theta_1) - ap$$

and

$$C_2 = c_2(Q_2 - a, \theta_2) - ap.$$

Uncertainty is modeled so that abatement costs of both countries are either fully independent or fully correlated. In reality, deviations from expectations about abatement costs in two countries would usually be partially correlated, so our modeling of uncertainty can be seen as boundary cases. Negative correlation is generally not to be expected, because "surprises" in abatement costs are likely to be positively correlated between different countries or regions.

To see this, consider typical reasons why actual abatement costs may deviate from ex-ante expectations. This may arise because technologies are less or more available or more or less costly than expected; because prices for energy and other production inputs differ from expectations; or because the behavior of businesses and individuals in response to emissions prices differs from that expected. In all of these cases, there is reason to expect that the surprise would occur and point in the same direction in all jurisdictions (or in both countries in our analysis), though the strength of its effect may be very different. Uncertainties that stem from policy making—for example, changes in policies that have interaction with emissions pricing—can be considered as largely independent between countries; again there is no reason to suspect that they would be negatively correlated.

In the default scenarios we assume the uncertainty parameters θ_i are log-normally distributed with a scale parameter of $\sigma = 0.4$, and we also test other parameter values. The effect of these scale parameters on the probability distributions for θ_i are shown in figure 6.2.

6.2 Application of the model

In this section, different sets of parameters are applied to the model by numerically evaluating expected costs for both countries under

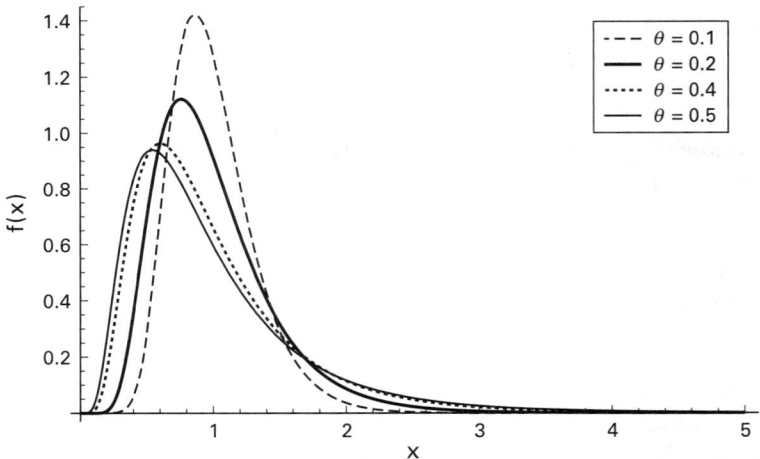

Figure 6.2
Example of log-normal distributions for different scale parameters σ to describe the probability distributions for the random variables θ_i as applied in the model.

uncertainty using the software package Wolfram Mathematica 8. In a first step, costs are evaluated assuming ex-ante identical countries, i.e., countries of same size and having the same parameters for uncertainty. The model application will focus on the effect of an additional tax introduced in country 1 on top of an existing cap-and-trade scheme on expected costs for both countries, and the distribution of abatement effort between the two countries. To do so, different levels of t_1 and overall ambition in producing the public good are considered.

In a second step, the model application is expanded from the case of identical countries to a case where parameters are chosen so that they mimic the situation in the EU ETS. While these countries are covered by a joint cap-and-trade scheme, the ambitions of climate and energy policies differ strongly between countries. Many EU member states introduced policies that aim to set incentives for greenhouse-gas reduction *in addition* to the EU ETS. Examples, as mentioned in the introduction of this chapter, include the UK carbon floor price, existing carbon taxes in several EU states, and the German renewable feed-in tariff scheme.

The objective of the approach taken here is not to generate empirically reliable figures on the effect of climate policy actions taken by EU member states, but to illustrate the partial effects of those policies on abatement and expected costs, in the situation where countries and sectors which are subject to the additional tax are not identical as in the simplified standard assumption. We do not claim to generate results which actually reflect the situation in the EU ETS because of the static partial equilibrium structure of the model, and choice of simple quadratic abatement cost functions (linear marginal abatement costs).

Ex-ante identical countries

Suppose that both countries are ex-ante identical, i.e., are of same size, have the same abatement costs ($\alpha_1 = \alpha_2 = 1$), and take on the same targets ($Q_1 = Q_2 = 1$). We compare expected costs under pure cap-and-trade without the additional tax ($t_1 = 0$) to the situation when country 1 levies an additional charge on top of the allowance price ($t_1 > 0$).

Figure 6.3 plots expected costs and amounts of emissions reductions as a function of the tax rate under three different assumptions about uncertainty: (1) that there is no uncertainty in the cost function; (2) uncertainty in cost is perfectly correlated for both countries; and (3) uncertainties in cost are independent.

As derived theoretically above, abatement costs increase for country 1 and decrease for country 2 as t_1 is increased. Under certainty, a corner solution will be reached at $t_1 = 2$. In this case, costs of country 2 become zero. Under certainty and $t_1 = 0$ (the reference case of pure cap-and-trade), the permit price would be 1. The tax rates in the figures below should be interpreted relative to this price. Abatement costs are lower when costs are correlated because individual realizations will include situations where different costs lead to gains from allowance trading.

Figure 6.3a also plots the expected amount of the public good that will be produced (solid/dashed black line). The overall quantity targets under cap-and-trade are set to equal $Q_1 + Q_2 = 2$ in this application. The amount of *expected* abatement (dashed black line) slowly increases as t_1 increases, because of the increasing probability of a corner solution with greater production of the public good than targeted.

Figure 6.4 shows independent and correlated expected total *unit cost* (or average cost) of abatement (sum of expected cost of both countries divided by expected abatement), as well as the expected allowance price and the expected number of traded allowances under correlated uncertainty. Unit cost and the number of traded allowances increase in t_1. The expected allowance price decreases as t_1 is increased. The elasticity of the allowance price under cap-and-trade with respect to the tax rate depends on relative abatement costs in both countries and the nature and extent of uncertainty.

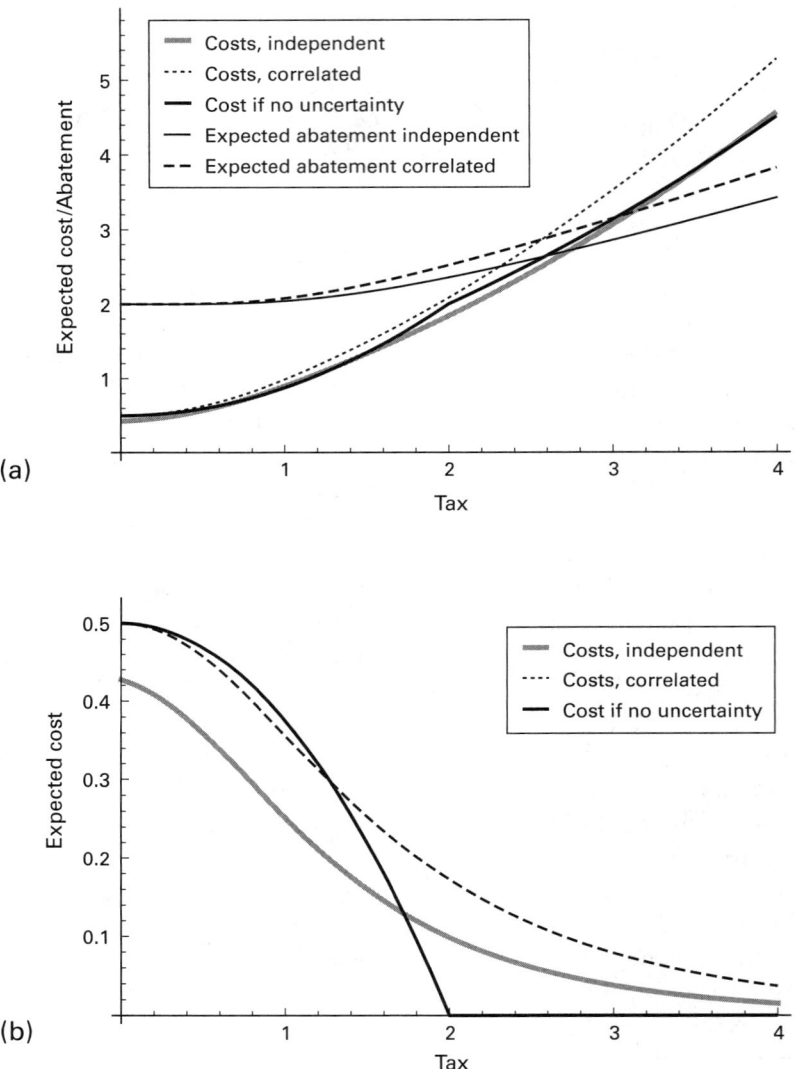

Figure 6.3
Expected costs for country 1 (panel a) and country 2 (panel b) as a function of the additional charge t_1 levied by country 1, with fully correlated and independent abatement costs, and assuming a scale parameter for uncertainty given by $\sigma = 0.4$. Panel a also shows expected abatement as a function of t_1.

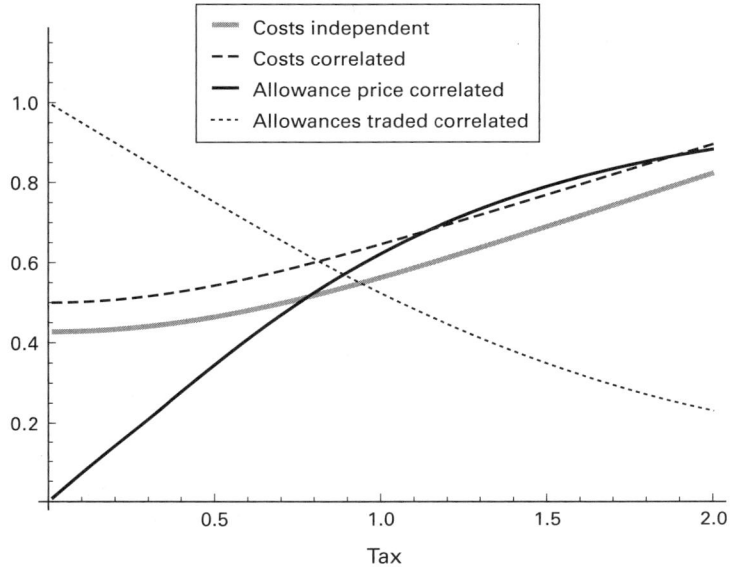

Figure 6.4
Expected independent and correlated total unit cost of abatement (average cost; sum of expected cost of both countries divided by expected abatement), expected correlated allowance price, and expected correlated allowances traded for ex-ante similar countries, assuming a scale parameter for uncertainty given by $\sigma = 0.4$.

The results are of particular interest with respect to the EU ETS. Countries that participate in the EU ETS are largely independent with respect to designing additional domestic climate policies and fiscal policy in general. In many EU countries additional policies for abatement beyond the EU ETS where introduced. However, countries are not ex-ante identical, but differ in terms of the size of regulated industries and abatement costs (for example through differences in technology). We will examine the case of the EU ETS in a stylized application to the EU's emissions trading scheme later in the chapter.

Stylized Application to the EU ETS
Under the burden-sharing agreement (EU 2002/358/EC), each EU country has been assigned some amount of allowable emissions (an emissions target), which implicitly leads to abatement obligations (Q_i in the model above). The developments in the EU in the first and second trading period of the EU ETS (2007–2012) have revealed that there is significant uncertainty about demand for allowances by

regulated companies and uncertainty about actual abatement costs (Hintermann 2010). Emissions regulated by the EU ETS dropped significantly since 2008, leading to a considerable surplus and dramatically lower allowance prices.

A number of EU countries have introduced or reinforced additional climate policies parallel to the EU ETS. An example is the UK carbon "price floor," which works as an additional levy on each tonne of CO_2 emitted by combustion installations regulated by the EU ETS.[6] The policy matches directly to our model, with the UK being the "tax country."

The UK combustion sector represents approximately 10 percent of total emissions covered under the EU ETS, so country 1 will be much smaller than country 2. On average, changes in emissions in the sector are positively correlated to the rest of the EU ETS, with a positive correlation of about 0.5.

Another example is the German renewable feed-in tariff scheme. While the policy offers subsidies for renewable energy instead of taxing greenhouse-gas emissions it still has an effect on the EU ETS by achieving some amount of emissions reductions in addition to that achieved by the EU ETS, and by thereby decreasing the allowance price within the ETS.

Since our model is a partial equilibrium model and uses only stylized assumptions about abatement costs and uncertainty, the results should not be interpreted as an actual evaluation of the effects of the UK price floor arrangement.[7] Our purpose is to give a broad indication about the magnitude of its effects, given in particular the relative magnitude of emissions covered by the "tax" and not covered.

To examine the effect of a "small country" (as part of the overall ETS) introducing an extra tax, we assume that the non-tax country 2 is ten times larger than country 1. We use a country size parameter $S = 10$, so that $Q_1 = 1$ and $Q_2 = SQ_1 = 10$, and adjust country 2's abatement cost parameter by α_2/S. We assume moderate uncertainty about abatement costs, with a scale parameter of $\sigma = 0.2$, and that abatement costs are perfectly correlated. Abatement costs in country 2 are slightly higher than in country 1 ($\alpha_1 = 1$, $\alpha_2 = 1.1/S$).

Table 6.1 and figure 6.5 show the allowance price under cap-and-trade; the number of traded allowances; abatement; and expected costs. If the tax were zero (pure cap-and-trade) the expected allowance price per unit of abatement would be 1.09. A small amount of allowances is sold from country 1 to country 2. Country 1's total costs are about 9

Table 6.1
Numerical evaluation of an additional tax under a joint cap-and-trade if country 2 is 10 times larger than country 1 ($S = 10$), if there is intermediate uncertainty about abatement costs ($\sigma = 0.2$), and country 2's abatement cost parameter is slightly higher than country 1's ($\alpha_1 = 1$, $\alpha_2 = 1.1$). Abatement costs are assumed to be perfectly correlated. All values (other than the tax rate) are expectations.

Tax	Price	Price change (%)	Tax/price ratio	Allowances traded	Allowances change (%)	Abatement	Cost 1	Cost 1 change (%)	Cost 2	Cost 2 change (%)	Total cost	Total cost change (%)
0.00	1.090	—	—	0.090	—	11	0.496	—	5.500	—	5.995	—
0.05	1.085	−0.45	0.046	0.144	60	11	0.498	0.39	5.499	−0.01	5.996	0.02
0.10	1.080	−0.91	0.093	0.198	120	11	0.503	1.38	5.498	−0.03	6.000	0.09
0.15	1.075	−1.36	0.140	0.252	180	11	0.511	2.97	5.497	−0.05	6.007	0.20
0.20	1.070	−1.82	0.187	0.306	240	11	0.521	5.15	5.496	−0.07	6.016	0.35
0.25	1.065	−2.27	0.235	0.360	300	11	0.535	7.94	5.494	−0.10	6.028	0.55
0.30	1.060	−2.73	0.283	0.414	360	11	0.552	11.32	5.492	−0.14	6.043	0.80
0.35	1.055	−3.18	0.332	0.468	420	11	0.572	15.30	5.490	−0.18	6.060	1.09
0.40	1.050	−3.64	0.381	0.523	480	11	0.595	19.89	5.487	−0.22	6.080	1.42
0.45	1.045	−4.09	0.430	0.577	540	11	0.620	25.07	5.485	−0.27	6.103	1.80
0.50	1.041	−4.55	0.481	0.631	600	11	0.649	30.85	5.482	−0.32	6.128	2.22
0.55	1.036	−5.00	0.531	0.685	660	11	0.681	37.23	5.478	−0.38	6.156	2.68
0.60	1.031	−5.45	0.582	0.739	720	11	0.715	44.21	5.475	−0.45	6.186	3.19
0.65	1.026	−5.91	0.634	0.793	780	11	0.753	51.78	5.471	−0.52	6.219	3.75
0.70	1.021	−6.36	0.686	0.847	840	11	0.793	59.96	5.467	−0.59	6.255	4.35
0.75	1.016	−6.82	0.738	0.901	900	11	0.837	68.73	5.463	−0.67	6.294	4.99
0.80	1.011	−7.27	0.791	0.955	960	11	0.883	78.11	5.458	−0.75	6.335	5.68
0.85	1.006	−7.73	0.845	1.009	1020	11	0.933	88.08	5.453	−0.84	6.379	6.41
0.90	1.001	−8.18	0.899	1.063	1080	11	0.985	98.65	5.448	−0.94	6.425	7.18
0.95	0.996	−8.64	0.954	1.117	1140	11	1.041	109.82	5.443	−1.03	6.475	8.00
1.00	0.991	−9.09	1.009	1.171	1200	11	1.099	121.59	5.437	−1.14	6.526	8.87

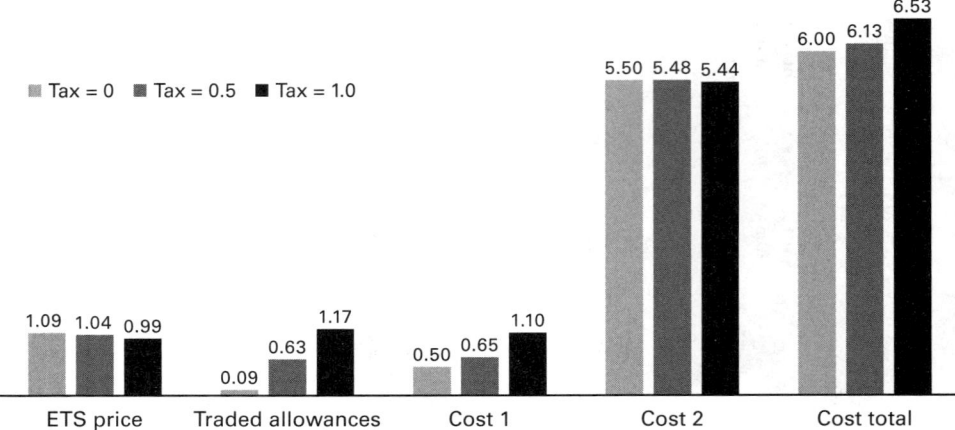

Figure 6.5
Expected values for ETS allowance price, number of traded allowances, cost for country 1 and country 2, and total cost for tax rates 0, 0.5, and 1 in country 1. (See table 6.1 for details.)

percent of country 2's costs. This is a least-cost situation where country 2 is large relative to country 1 and has higher abatement costs.

As country 1 introduces a tax, the price of allowances decreases. At tax rate $t = 0.55$ (about 50 percent of the allowance price) the expected price is about 5 percent less than it is when $t_1 = 0$. If country 1 sets the tax rate approximately equal to the cap-and-trade price, allowance prices decrease by about 9 percent relative to $t_1 = 0$. Since country 1 is considerably smaller than country 2, the effect of the tax on the allowance price is relatively small. The amount of allowances sold from country 1 to country 2 increases as t is increased. This is because a greater share of the overall abatement is produced in country 1 on account of the tax.

Since more abatement occurs in country 1 for $t_1 = 0$, the total abatement costs of country 1 increase as t_1 increases. While the feedback of the tax on the allowance price is moderate if country 1 is small, cost increases in country 1 are considerable. For a tax of about 50 percent of the allowance price, costs for country 1 increase by about 37 percent relative to $t_1 = 0$. If the tax is set to approximately equal to the allowance price ($t_1 = 1$) country 1's domestic abatement costs will be more than twice as much as they are when $t_1 = 0$. Cost increases in country 1 are accompanied by small cost decreases in country 2. This is because country 2 can purchase allowances from country 1 and allowance prices decrease as t_1 is increased. Combined costs increase relative to the situation without a tax because country 1's cost increases cannot be offset by cost decreases in country 2.

The model application showed that the introduction of an additional tax under cap-and-trade and uncertain abatement costs will have a moderate effect on prices and total abatement costs if the country or sector covered by the tax is small relative to the remaining countries or sectors, and if the tax rate is small relative to the allowance price. For example, a tax rate of 10 percent of the allowance price in the model application leads to a decrease in the allowance price of about 1 percent and an increase in total costs of about 0.1 percent. If country 1 is small, it further is unlikely that a corner solution will be reached (all of the combined abatement task fulfilled by the tax in country 1 alone).

Allowance price elasticity, country size, and uncertainty
In this subsection we examine the elasticity of the cap-and-trade price with respect to the tax rate, and its relationship to country size and uncertainty. Under the parameterization of this chapter, elasticities are not constant. This approach helps provide an understanding of how prices are affected by country size and correlations in uncertainty parameters. The elasticity of the cap-and-trade price with respect to the tax thus gives an indication of efficiency implications of the additional tax. In our simple model (omitting effects of innovation and other long-run effects), the additional tax always leads to additional costs for the country that levies the tax in addition to the cap-and-trade scheme. The actual extent of additional costs, however, depends on the feedback of the unilateral tax on the joint cap-and-trade scheme, i.e., the elasticity of the allowance price with respect to the tax.

We calculate expected prices by numerical integration[8] for tuples of $S = 1, 2, \ldots, 10$ and $\sigma = 0.05, 0.1, \ldots, 0.5$. For each S,σ tuple, prices are evaluated for $t = 0.1, 0.11, \ldots, 1$ so that price-tax combinations are obtained for each of the S,σ tuples. To estimate the average elasticity of cap-and-trade prices with respect to the tax we consider a simple linear elasticity model of the form

$\log(price_i) = \beta_0 + \beta_1 \log(tax_i) + u_i.$

The linear elasticity model is estimated for S,σ tuples to obtain elasticity $\beta_{1, S,\sigma}$. If the tax in country 1 increases by 1 percent, the price in the joint cap-and-trade scheme changes by $\beta_{1, S,\sigma}$ percent.

Tables 6.2 and 6.3 show elasticity $\beta_{1, S,\sigma}$ for independent and correlated uncertainty. The elasticity of the price under cap-and-trade with

Table 6.2
Elasticity of the allowance price for changes in the tax rate for different country size (S) and uncertainty scale parameter (σ). Cost uncertainty is independent.

S	σ 0.05	0.1	0.15	0.2	0.25	0.3	0.35	0.4	0.45	0.5
1	−0.30082	−0.31017	−0.31804	−0.32439	−0.32945	−0.33349	−0.33674	−0.33936	−0.34149	−0.34323
2	−0.17815	−0.18478	−0.19114	−0.19716	−0.20276	−0.20791	−0.21263	−0.21693	−0.22085	−0.22442
3	−0.12684	−0.13185	−0.13668	−0.14136	−0.14586	−0.15015	−0.15423	−0.15809	−0.16173	−0.16516
4	−0.09854	−0.10256	−0.10646	−0.11025	−0.11392	−0.11747	−0.12089	−0.12418	−0.12734	−0.13036
5	−0.08058	−0.08395	−0.08722	−0.0904	−0.09349	−0.0965	−0.09942	−0.10225	−0.10499	−0.10763
6	−0.06816	−0.07106	−0.07389	−0.07663	−0.07931	−0.08191	−0.08445	−0.08692	−0.08933	−0.09166
7	−0.05907	−0.06161	−0.0641	−0.06652	−0.06887	−0.07118	−0.07342	−0.07562	−0.07775	−0.07984
8	−0.05211	−0.05438	−0.0566	−0.05876	−0.06088	−0.06294	−0.06496	−0.06693	−0.06885	−0.07073
9	−0.04663	−0.04867	−0.05068	−0.05263	−0.05455	−0.05642	−0.05825	−0.06004	−0.06179	−0.0635
10	−0.04219	−0.04405	−0.04588	−0.04767	−0.04941	−0.05113	−0.0528	−0.05444	−0.05605	−0.05762

Table 6.3
Elasticity of the allowance price for changes in the tax rate for different country size (S) and uncertainty scale parameter (σ). Cost uncertainty is perfectly correlated.

S	σ 0.05	0.1	0.15	0.2	0.25	0.3	0.35	0.4	0.45	0.5
1	-0.29073	-0.29041	-0.28934	-0.28757	-0.28531	-0.28274	-0.27999	-0.27717	-0.27433	-0.27154
2	-0.17129	-0.17128	-0.17126	-0.17116	-0.17096	-0.17065	-0.17024	-0.16974	-0.16918	-0.16857
3	-0.12166	-0.12166	-0.12166	-0.12165	-0.12163	-0.12158	-0.12150	-0.12139	-0.12126	-0.12110
4	-0.09438	-0.09438	-0.09438	-0.09438	-0.09437	-0.09437	-0.09435	-0.09432	-0.09429	-0.09425
5	-0.07711	-0.07711	-0.07711	-0.07711	-0.07710	-0.07710	-0.07710	-0.07710	-0.07709	-0.07708
6	-0.06518	-0.06518	-0.06518	-0.06518	-0.06518	-0.06518	-0.06518	-0.06518	-0.06519	-0.06519
7	-0.05646	-0.05646	-0.05646	-0.05646	-0.05646	-0.05646	-0.05646	-0.05646	-0.05646	-0.05647
8	-0.04979	-0.04979	-0.04979	-0.04979	-0.04979	-0.04979	-0.04979	-0.04980	-0.04980	-0.04981
9	-0.04453	-0.04453	-0.04453	-0.04453	-0.04453	-0.04453	-0.04454	-0.04454	-0.04454	-0.04455
10	-0.04028	-0.04028	-0.04028	-0.04028	-0.04028	-0.04028	-0.04028	-0.04029	-0.04029	-0.04030

respect to the tax is dependent on relative country size S and on the uncertainty scale parameter σ. For independent abatement costs, a larger uncertainty scale parameter leads to higher sensitivity of the cap-and-trade price with respect to changes in the tax rate. If country 1 is small, the effect of changes in the tax rate tend to be relatively low. For $S = 10$ and $\sigma = 0.2$, a 1 percent increase in the tax rate would lead to an average decrease of 0.048 percent in the cap-and-trade price.

For perfectly correlated abatement costs, the elasticity of the price with respect to the tax is in general lower than when abatement costs are independent, while the effect of uncertainty is reversed. The elasticity is slightly lower for higher values of the scale parameter σ. In particular for the case of country 1 being relatively large both countries can encounter high cost if there is perfectly correlated uncertainty. Therefore, the effect of the tax on the price can be less pronounced for larger values of σ, i.e., since the tax rate was assumed not to exceed $t = 1$ in this application. With respect to country size, the previous results also hold in the case of correlated uncertainty, where the elasticity of the price with respect to the tax is rather low if S is large and country 1 is small relative to country 2.

For a case like the UK power sector, it might be reasonable to assume $S = 10$ with uncertainty and some positive correlation (not necessarily perfect correlation). As tables 6.2 and 6.3 show, there is not much difference between the cases of perfectly correlated and uncorrelated abatement costs. The model results indicate that the elasticity of the cap-and-trade price with respect to the tax (for $t = 0.1, 0.11, \dots, 1$) will be between −0.040 and −0.058 if a "small country" introduces a tax.

6.3 Discussion and consideration of policy implications

Our stochastic partial equilibrium model demonstrates that unilateral additional efforts for public good production (i.e., an additional unilateral tax) under a joint cap-and-trade scheme will cause excess costs and will in most cases not yield additional production of the public good. Two cases can be distinguished.

In the interior solution the cap-and-trade price will be greater than zero in the presence of the tax. In this situation more abatement efforts are borne by country 1. Costs for country 1 increase accordingly, while costs for country 2 decrease. Total costs (the sum of cost of both countries) will increase since increased costs of country 1 cannot be offset by cost savings in country 2. There is no additional production of the public good (abatement) in the interior solution since the joint quantity target $Q_1 + Q_2$ is binding.

In the corner solution, all abatement is borne by country 1 and there is no need for further abatement by country 2. The allowance price is zero in the corner solution.

Expected costs for country 1 and expected total costs will in most cases be considerably higher than with pure cap-and-trade without the additional tax. The allowance price decreases, and extra costs increase, as the tax rate is increased.

The stochastic modeling showed that country size and the magnitude of uncertainty about abatement costs are important determinants of the effect of the tax on the allowance price. In section 6.2 we examined the case of a small country with additional tax. The application showed that the introduction of the tax will lead to increased expected allowance sales from country 1 to country 2 (see table 6.1), so that public good production (and cost) is re-allocated from country 2 to country 1. Expected costs for country 1 increase considerably, while the cost of country 2 decrease slightly, and combined costs increase due to the divergence in marginal costs.

The elasticity of the allowance price with respect to the tax for different country size and levels of uncertainty was examined in section 6.2. For the chosen parameter set, the effect of country size on the elasticity is more pronounced than for the level of uncertainty. Expected allowance prices decrease as the tax is increased. If country 1 is small, the effect of the tax on the allowance price is moderate.

One reason for country 1 to introduce an additional tax could be that the country aims to speed up the decarburization of the economy and to set stronger incentives for innovation (relative to country 2). Such aspects are omitted in our model, but may be of importance in theory and practice, if stronger action is viewed as improving dynamic efficiency or desired for other reasons. Carbon pricing can offer strong incentives for innovation, and again "country size" is an important determinant of the prospects for reaping dynamic efficiency gains (Milliman and Prince 1989; Downing and White 1986). The "stringency" of regulation can be important for the extent of innovation, as shown for the EU ETS and the power sector by Rogge et al. (2011). Innovation research in the field of environmental economics has further emphasized the dual aspect of environmental regulation on externalities. While negative externalities, such as pollution, are decreased, the incentives for innovation can be interpreted as a positive externality caused by a price (Rennings 2000).

Our model does not account for the effects of innovation. However, for additional policies in one country to be globally welfare enhancing, the positive innovation externality must at least outweigh excess costs of the policy. Innovation activity can be expected to increase in the country that introduces the additional policy, but could decrease in the other country on account of the lower allowance price.[9] This could lead to overall greater innovation activity, for example if the greater incentives in the country with the additional policy allow the crossing of an "innovation threshold" that would not be reached under a pure cap-and-trade scheme. Whether and under which conditions this may hold is an empirical question beyond the scope of this chapter.

Conclusion

We have examined the introduction of additional national policies for public good production, specifically a national carbon tax in one country, under a joint cap-and-trade scheme of two countries with fixed joint quantity targets. Examples for such additional policies would be the UK carbon price floor or the German renewable feed-in tariff scheme in combination with environmental regulation by the EU Emissions Trading Scheme. Both unilateral policies aim at the utility sector, which already is regulated by the EU ETS.

In most cases the additional tax will not lead to additional abatement beyond the fixed quantity target. Additional abatement will only occur in the case when a corner solution is reached. In this case all abatement is undertaken by the country with the extra tax, the allowance price under cap-and-trade will be zero, and expected costs are considerably higher.

Countries may pursue dynamic efficiency objectives in strengthening their domestic effort, including through fostering the innovation effort. We do not evaluate this aspect, rather we investigate the effect of relative country size and magnitude of abatement cost uncertainty.

The expected allowance price decrease as the tax rate is increased. The effect depends on the relative size of the country that introduces the additional policy and the extent of uncertainty. The smaller the country with additional policy is, the smaller will be the feedback of the tax on the allowance price. Larger uncertainty will lead to a stronger feedback effect if abatement costs are independent and will lead to smaller feedback effect if costs are perfectly correlated.

Our model results highlight the importance of interactions between domestic climate policies and overarching mechanisms such the EU ETS. Countries may have specific national objectives in introducing additional policy mechanisms domestically. However such additional national policies will generally not result in overall increased effort, and will result in higher overall static abatement cost. It is possible that there are gains in dynamic efficiency, for example through enhanced innovation effort in the country with the additional policies. Whether and under which conditions these exist and outweigh the additional static costs that arise from diverging marginal costs of abatement remains an open question.

Acknowledgments

We are grateful for funding by the Deutscher Akademischer Auslandsdienst (DAAD) and the Group of Eight (Go8) through the Go8 Joint Research Cooperation Scheme. The research has further received funding from the European Community's Seventh Framework Programme under Grant Agreement No. 308481 and also from the Australian Research Council (DP110102057). Such support does not imply agreement with the views expressed in this chapter. We are grateful for comments by Timothy Fitzgerald and participants at the CESifo Summer Institute 2013 "Emissions Trading Systems as a Climate Policy Instrument," organized by Marc Gronwald and Beat Hintermann.

Notes

1. Modeling uncertainty under price and quantity-based emissions control as a random change in the slope of the marginal abatement cost curve is in the tradition of Weitzman 1974 and the subsequent literature, and we follow this convention. It would also be possible to model uncertainty about underlying emissions growth and thus the amount of effort required to meet any chosen emissions target, but this would complicate the analysis without providing additional insights.

2. Transaction costs (Stavins 1995), market power (Hahn 1984), and non-cost-minimizing behavior (Hahn and Stavins 2011) are distortions under which marginal abatement costs might not be equated and a least-cost solution might not be achieved.

3. This is the key difference between the model we use here and the one used in Wood et al. 2013, where it is assumed that country 1 only levies a tax and the price from the cap-and-trade scheme applied only in country 2.

4. Assuming compliance with the quantity targets is equivalent to assuming a "very high" compliance penalty in the emissions trading scheme in country 2.

5. Marginal abatement cost curves can be non-differentiable as shown by McKitrick (1999) and can have high inter-sectoral and intra-sectoral variance, as shown for the case of different air pollutants by Hartman et al. (1997), in which case simplifying assumptions are needed.

6. A "price floor" normally describes a minimum price within an ETS, implemented for example through a reserve price at auction (Wood and Jotzo 2011). The UK scheme arrangement is related but different to this standard version of a price floor.

7. We assume quadratic abatement cost functions for both countries and a log-linear probability distribution to calculate expected values by numerical integration using the software package Wolfram Mathematica. For a detailed description of the parameterized model and the structure of uncertainty please see (Wood et al. 2013, pp. 4–7).

8. For detailed information on how uncertainty is modeled, see pp. 5–7 of Wood et al. 2013.

9. This argument is related to dynamic efficiency of regulation discussed in Downing and White 1986.

References

Baumol, W. J. 1972. On taxation and the control of externalities. *American Economic Review* 62 (3): 307–322.

Baumol, W. J., and W. E. Oates. 1971. The use of standards and prices for protection of the environment. *Swedish Journal of Economics* 73 (1): 42–54.

Baumol, W. J., and W. E. Oates. 1988. *The Theory of Environmental Policy*, second edition. Cambridge University Press.

Boeters, S., and J. Koornneef. 2011. Supply of renewable energy sources and the cost of EU climate policy. *Energy Economics* 33 (5): 1024–1034.

Böhringer, C., A. Löschel, U. Moslener, and T. F. Rutherford. 2009. EU climate policy up to 2020: An economic impact assessment. *Energy Economics* 31: 295–305.

Böhringer, C., T. F. Rutherford, and R. S. J. Tol. 2009. The EU 20/20/2020 targets: An overview of the EMF22 assessment. *Energy Economics* 31: 268–273.

Bovenberg, A. L. 1999. Green tax reforms and the double dividend: An updated reader's guide. *International Tax and Public Finance* 6: 421–443.

Bovenberg, A. L., and L. H. Goulder. 1996. Optimal environmental taxation in the presence of other taxes: General-equilibrium analyses. *American Economic Review* 86 (4): 985–1000.

Bovenberg, A. L., and F. van der Ploeg. 1994. Environmental policy, public finance and the labour market in a second-best world. *Journal of Public Economics* 55: 349–390.

Downing, P. B., and J. L. White. 1986. Innovation in pollution control. *Journal of Environmental Economics and Management* 13 (1): 18–29.

EU 2002/358/EC. 2012. Council Decision of 25 April 2002 concerning the approval, on behalf of the European Community, of the Kyoto Protocol to the United Nations Framework Convention on Climate Change and the joint fulfilment of commitments thereunder.

Hagem, C., and H. Westskog. 1998. The design of a dynamic tradeable quota system under market imperfections. *Journal of Environmental Economics and Management* 36: 89–107.

Hahn, R. W. 1984. Market power and transferable property rights. *Quarterly Journal of Economics* 99 (4): 753–765.

Hahn, R. W., and R. N. Stavins. 2011. The effect of allowance allocations on cap-and-trade system performance. *Journal of Law & Economics* 54 (4): 267–294.

Hartman, R. S., D. Wheeler, and M. Singh. 1997. The cost of air pollution abatement. *Applied Economics* 29 (6): 759–774.

Hintermann, B. 2010. Allowance price drivers in the first phase of the EU ETS. *Journal of Environmental Economics and Management* 59 (1): 43–56.

IEA. 2013. *Energy Policies of IEA Countries*. Germany: Review.

McKitrick, R. 1999. A derivation of the marginal abatement cost curve. *Journal of Environmental Economics and Management* 37 (3): 306–314.

Milliman, S., and R. Prince. 1989. Firm incentives to promote technological change in pollution control. *Journal of Environmental Economics and Management* 17 (3): 247–265.

Montgomery, W. D. 1972. Markets in licenses and efficient pollution control programs. *Journal of Economic Theory* 5: 395–418.

OECD. 2013. *Taxing Energy Use: A Graphical Analysis*.

Pezzey, J. J. C. V., and F. Jotzo. 2012. Tax-versus-trading and efficient revenue recycling as issues for greenhouse gas abatement. *Journal of Environmental Economics and Management* 64 (2): 230–236.

Rennings, K. 2000. Redefining innovation—eco-innovation research and the contribution from ecological economics. *Ecological Economics* 32 (2): 319–332.

Rogge, K. S., M. Schneider, and V. H. Hoffmann. 2011. The innovation impact of the EU Emission Trading System: Findings of company case studies in the German power sector. *Ecological Economics* 70 (3): 513–523.

Sinn, H.-W., and U. Schmoltzi. 1981. Eigentumsrechte, Kompensationsregeln und Marktmacht: Anmerkungen zum Coase Theorem. *Jahrbücher für Nationalökonomie und Statistik* 196 (2): 97–117.

Stavins, R. N. 1995. Transaction costs and tradeable permits. *Journal of Environmental Economics and Management* 29 (2): 133–148.

Toke, D. 2011. UK electricity market reform—Revolution or much ado about nothing? *Energy Policy* 39 (12): 7609–7611.

Weitzman, M. L. 1974. Prices vs. quantities. *Review of Economic Studies* 41:477–491.

Wood, P. J., P. Heindl, F. Jotzo, and A. Löschel. 2013. Linking price and quantity pollution controls under uncertainty. Working paper 1302, Centre for Climate Economics and Policy, Crawford School of Public Policy, Australian National University.

Wood, P. J., and F. Jotzo. 2011. Price floors for emissions trading. *Energy Policy* 39 (3): 1746–1753.

7 Interaction between EU Instruments and Member-State Instruments: The End of CO_2 Emissions Trading in Europe?

A. J. Mulder

Member states of the European Union have chosen to cap CO_2 emissions from energy-intensive sectors via the European Union Emissions Trading Scheme to ensure that collective long-term emissions-reduction goals are achieved. Theoretically, the market-based character of the EU ETS should ensure that the emission target is achieved in a least-cost manner. However, such a least-cost solution is possible only if outside interference with the allocation via the EU ETS market scheme is avoided (Böhringer, Koschel, and Moslener 2008). Interestingly, European policy makers themselves are likely to be a source of outside interference by introducing many instruments for CO_2 abatement on a national level alongside the EU ETS. Many of those parallel instruments are introduced in pursuit of domestic energy and climate targets (EEA 2011; Lundberg et al. 2012). Examples include power-plant performance benchmarks, feed-in-tariffs for renewables, and biomass co-firing mandates. Parallel instruments can have local benefits for the national government that introduces the instrument, such as employment benefits or stability of electricity supply (Sorrell and Sijm 2003; Bennear and Stavins 2007). However, with respect to carbon abatement, parallel instruments are direct substitutes for the EU ETS. The abatement achieved through parallel instruments reduces the demand for EU ETS carbon allowances and lowers the carbon price, thereby reducing the amount of abatement that is triggered elsewhere in Europe by the EU ETS. Building on this logic, the aggregate impact of all parallel instruments across Europe could significantly lower the carbon price and increase the societal costs associated with CO_2 abatement. At an extreme, parallel instruments could make the EU ETS completely redundant, permanently driving the CO_2 allowance price down to zero.

Aside from the fact that national governments have introduced a wide variety of parallel instruments in recent years, the EU ETS carbon price has been rather low and volatile. The current weak performance of the EU ETS is typically attributed to negative and/or stagnating economic growth since 2008, in combination with a too generous allowance allocation regime. The potential role of parallel instruments is not well understood in this context, yet such knowledge is needed in order to formulate the right policy response if policy makers are interested in strengthening the EU ETS. Also, greater knowledge of the effects of multiple parallel instruments on the performance of the EU ETS can be part of the solution. As long as policy makers are unaware of the costs associated with parallel instruments in the form of reduced ETS performance, they are inclined to spend more than the socially optimal amount on parallel instruments. Closing this information gap could be an important step toward a more cost-efficient and goal-oriented policy design.

Although a deep understanding of the effect of parallel instruments on the performance of an emissions trading scheme can be highly valuable to policy makers, the focus of existing literature has been largely limited to interactions between an ETS and a single other parallel instrument. Studying the aggregate effect of multiple parallel instruments would allow for a more comprehensive sensitivity analysis of the EU ETS and would represent a more realistic policy setting. In this chapter, I intend to add to the literature by empirically examining the performance of the EU ETS within a policy setting with multiple parallel instruments. As will be explained in more detail below, I aim to provide benchmarks to policy makers that can be used to assess the potential adverse effect of a proposed parallel instrument on the performance of the EU ETS.

In my analysis, I distinguish between two broad categories of parallel instruments: Type 1 and Type 2 instruments. Both types lead to a reduction of emissions in ETS sectors, but they do so in different ways. Type 1 parallel instruments are defined as instruments that provide *ETS sectors* with incentives to adopt low-carbon technology. Thereby, Type 1 incentives lower the carbon intensity of production in ETS sectors. An example of a Type 1 instrument is a subsidy to invest in biomass co-firing in existing coal-fired power plants. Type 2 instruments are defined as instruments that provide incentives to *non-ETS sectors* (e.g., households) to lower their demand for products from ETS sectors. In that manner, Type 2 instruments lower the required production level in ETS sectors and the associated CO_2 emissions. Two examples of Type 2 instruments are incentives for deployment of decentralized renewable electricity generation capacity and subsidies to improve the energy efficiency of households. Both Type 1 and Type 2 instruments have been introduced on a relatively large scale across Europe. A few examples of both types of instruments that are currently in force are provided in table 7.1. (All the examples in the table have been introduced since 2008.)

To study the effects of both Type 1 and Type 2 instruments on indicators of the performance of the EU ETS, I use a stochastic simulation model of the EU ETS (Mulder and Jepma 2015) that incorporates year-to-year economic growth uncertainty. The performance of the EU ETS is known to be highly dependent on economic growth rates: high economic growth rates force firms to invest heavily in CO_2 abatement to remain below the CO_2 allowance cap and provide upward pressure for the carbon price, whereas the carbon price and the need to invest in CO_2 abatement is significantly lower if economic growth rates fall below average. Incorporating both parallel instruments and economic growth uncertainty in the analysis makes it possible to examine their relative importance.

I put particular emphasis on the conditions for the EU ETS to become redundant because its redundancy provides a clear benchmark for policy makers. EU ETS redundancy is defined as a situation in which parallel instruments trigger enough abatement activity (in million metric tons of carbon dioxide per year, abbreviated $MtCO_2/yr$) to permanently drive the EU ETS carbon price to zero. If it is assumed that policy makers have knowledge about the expected local abatement effects of a proposed parallel instrument (in $MtCO_2/yr$), the threshold level enables policy makers to assess the relative EU-ETS-undermining effect of the proposed parallel instrument. In that manner, policy makers are better informed about potential costs of national policy development alongside the EU ETS.

7.1 Literature review

The study of interacting policy instruments has its roots in the work of Tinbergen (1952, 1956), who formulated general rules for the controllability of an economic system. He coined the rule that the number of independent instruments must equal the number of independent targets in order for a solution to exist. Theil (1954, 1956, 1964) extended the work of Tinbergen to other situations, including situations in which the number of instruments is lower than the number of targets. My

Table 7.1
Examples of Type 1 and Type 2 instruments that reduce ETS sector emissions since 2008 across Europe.

Type 1	Instruments aimed at ETS sectors: reduce carbon intensity of production in ETS sectors	Type 2	Instruments aimed at non-ETS sectors: reduce production levels in ETS sectors
Austria (2012)	Ökostromverordnung—FITs for biomass co-firing (€0.0612/kWh).	Austria (2012)	Ökostromverordnung—FITs for renewable energy.
Netherlands (2009)	Agreement on energy efficiency for ETS companies (MEE)—Negotiated agreement that forces ETS firm to aim for energy efficiency improvement.	Netherlands (2011)	SDE+—Provides a feed-in subsidy to installations according to generation costs on a first come first served basis.
Germany (2012)	CHP Agreement with Industry—Agreement between German Government and the industrial sector to improve energy efficiency in the industrial sector. Objective: raise energy efficiency by 1.3% annually.	Germany (2011/2012)	2011—Energy Efficiency Fund—Fund of >€100 million to promote energy efficiency across end-use sectors. 2012—Up to 30% financial allowance for investments in cross-sectional technology that increases energy efficiency (e.g. heat pumps and air conditioning). 2012—Amendment of EEG—Feed-in-Tariffs (FITs) for non-ETS renewables.
Italy (2008)	Decree on Implementation of EU Energy Services Directive—Includes setting up a White Certificate Scheme in the energy industry.	Italy (2012)	Ministerial Decree—Incentives for increased energy efficiency in existing buildings, totaling €200 million in subsidies
Poland (2008)	Long-term Programme for Promotion of Biofuels or Other Renewable Fuels—Provides support for biomass co-firing through arrangements that improve cost-effectiveness of biomass supply chain.	Poland (2011)	Energy Efficiency Act—Introduces a White Certificate Scheme imposed in utility companies that promotes energy efficient behavior of customers.

Portugal (2008/2010)	2008—Management System of Intensive Energy Consumption—Binding energy audits for energy-intensive facilities (>500 tonnes of oil equivalent per year) with a 6–8 year interval. Facility operators have to set energy and carbon intensity targets. After approval by government, penalties can be issued for missing the target. 2010—Implementation of CHP Directive—Provides financial remuneration for high efficiency and renewable based electricity generation in CHP plants.	Portugal (2010/2013)	2010—Tax Deduction for Efficient Equipment—Tax deductions on investments in efficient equipment that improve the thermal performance of buildings. 2013—Feed-in tariffs for micro and mini generation for 2013—Includes feed-in tariff for mini (<3.68kW) and micro (3.68–20 kW) solar PV for 15 years: first 8 years €0.196/kWh, following 7 years €0.165/kWh.
Spain (2008)	Voluntary Agreements 2008–2012—Promotes adoption of energy saving measures by industry. Financing lines are available, with preferential treatment for formally committed firms.	Spain (2013)	PIMA SOL—program to promote GHG reduction in the tourism sectors via, amongst others, reduced energy consumption.
Sweden (2010)	Energy Audit for Companies—provides support for 50% of costs of an energy audit for companies using more than 500 MWh/yr Measures follow a few years later.	Sweden (2010)	Government subsidies for Local Energy Efficiency Measures— ~€11 million annually for local municipalities and county councils to undertake energy efficiency measures.
UK (2010)	National Renewable Energy Action Plan—Includes Renewable Obligation Certificates (ROC) to subsidize biomass-co-firing.	UK (2010)	CRC Energy Efficiency Scheme—Targets large private and public sector organizations and caps their emissions.

source: IEA Policies and Measures Database (http://www.iea.org/policiesandmeasures/)

analysis in this chapter concerns the reverse situation: an over-determined system with more instruments than targets. On top of that, multiple governments govern the instruments, while the EU ETS is an instrument shared by all thirty-one governments. Over-determined systems have many solutions, although such a solution may be hard, if not impossible, to attain in practice. Finding a solution requires strong coordination by a central planner. That planner should set all excess instruments at arbitrary fixed values, while having full information regarding the relations (or lack thereof) among instruments, targets, and the behavior of the private sector. Without a social planner or full information, a solution becomes indeterminate and uncontrollable for all governments involved (Acocella, Di Bartolomeo, and Hughes Hallett 2012; Di Bartolomeo, Hughes Hallett, and Acocella 2011).

A set of interacting instruments and targets can become so complex that retaining control over them becomes a cause in itself for policy makers (Wildavsky 1979). Majone (1989) defines *policy space* as meaning a set of policies that are so closely interrelated that it is not possible to give useful descriptions of one of them or to make analytic statements about one of them without taking the others into account. Majone builds on Wildavsky's work by pointing out that policy makers tend to lose control over the policy space over time. As the number of policies grows relative to the size of the policy space, policies logically become more interdependent and interfere with other policies. At an extreme, new programs and institutional arrangements may be required to prevent or reduce the unwanted consequences of a congested policy space.

Bye and Bruvoll (2008) suggest that policy development in the energy and environmental domain has already resulted into an over-congested policy space. Concluding that little is known about the aggregate effect of environmental instruments, they call for coordination and simplification of policy tools before new instruments are added to the policy space.

To the extent that policy interactions have been analyzed within the environmental domain, the primary focus has been on interactions between an emissions trading scheme and a scheme that supports the deployment of renewable electricity technologies. Much of this work has focused on the expected changes in the prices (Boots 2003; Rathmann 2007) and in the supply of electricity (Anandarajah and Strachan 2010), on welfare implications (Böhringer et al. 2008), on the CO_2 price

(Hindsberger et al. 2003), and on levels of CO_2 emissions (Morthorst 2003). (See Del Rio 2007 for another review of the literature.) Fewer authors have considered highly congested policy spaces, although such analyses could help to uncover and avoid unintentional consequences of congestion. In what follows I will highlight some notable papers in which a more congested policy space has been considered.

Oikonomou and Jepma (2008) designed a qualitative framework to identify potential interactions between combinations of various climate policy instruments. Their framework builds on, and summarizes, interactions that have been identified in previous literature, departing from findings of the INTERACT project (Sorrell 2003). The framework helps policy makers to classify potential positive and negative interactions between sets of instruments. Kautto et al. (2012), using the existing literature and interviews with experts, analyzed changes in the use of biomass as a result of the introduction of the EU ETS and its interactions with parallel instruments in seven EU countries. Although Kautto et al. had difficulty attributing observed effects to specific instruments, they noted that the EU ETS probably had amplified the effects of existing policies. In some cases the introduction of the EU ETS triggered the introduction of additional "balancing measures" to offset biomass price effects. Sorrell and Sijm (2003) and Bennear and Stavins (2007) identified situations in which combinations of environmental instruments can be justified. Sorrell and Sijm (2003) argued that combinations of instruments can usefully coexist if they lead to an improvement of the static of dynamic efficiency of a trading scheme or if they deliver other valuable policy objectives. Bennear and Stavins (2007) noted that multiple instruments can be justified if there have been multiple market failures or if an exogenous constraint cannot be removed.

Although the use of multiple instruments can have benefits, it remains unclear when such combinations can lead to a loss of control by policy makers. However, one can safely assume that instruments are never introduced by policy makers with the intent to be redundant. If instruments become redundant unintentionally, it indicates that policy makers have lost control over the policy space, as Majone (1989) and Wildavsky (1979) suggested. De Jonghe et al. (2009) analyzed the possibility of redundancy of an emissions trading scheme, albeit in a stylized theoretical setting with one parallel instrument. Employing a welfare-optimization model, they showed that if a renewables quota[1] is set above a threshold level alongside an emissions trading scheme, the CO_2 allowance price falls to zero and the ETS becomes redundant.

These results, in line with what Hindsberger et al. (2003) have found,[2] seem to suggest that if policy makers set their renewables quota below a threshold level, the ETS will produce a positive carbon price and contribute to carbon abatement. However, that this is not necessarily true in an international setting. That is, even if a national government sets a relatively low renewables quota, actions by other national governments across Europe could still drive the EU ETS beyond the threshold level and into redundancy. In fact, the introduction of a renewables quota in one country could actively trigger other governments to implement additional instruments in response to the depreciated carbon price. If such dynamics between policy-making authorities are disregarded, the probability that an ETS becomes redundant is therefore likely to be underestimated.

The likelihood of EU ETS redundancy is even greater once we take multiple sectors into account. The EU ETS covers many countries and industrial sectors and therefore interacts with a wide range of energy-related and climate-related instruments. An ETS even interacts with instruments in sectors that are not covered by the scheme (Interact, 2003). Despite these facts, literature in the field (Morthorst 2001, 2003; Del Rio 2009; Hindsberger et al. 2003; Conrad and Kohn 1996; Rathmann 2007; Amundsen and Mortensen 2001; Jensen and Skytte 2003) often focuses exclusively on the power sector.

General-equilibrium models, which cover multiple sectors, typically model an emissions trading scheme that is too simple to fully assess the adverse effects of parallel instruments (Morris, Reilly, and Paltsev 2010; Abrell and Weigt 2008; Pizer 2002). Allowance banking behavior and the stochastic nature of both economic growth and CO_2 allowance demand are typically not accounted for. These factors are, however, rather important to obtain a full understanding of the effect of introducing parallel instruments alongside an emissions trading scheme (Rathmann 2007). In this chapter, I take the above-mentioned factors into account. That is, I analyze policy interaction in an international setting, with parallel instruments in both ETS and non-ETS sectors. Also, I apply a model that accounts for important design features of the EU ETS, such as the ability to bank allowances and the role of stochastic allowance demand patterns. The model reflects current EU ETS regulation. I aim to provide a more complete analysis of the sensitivity of the EU ETS to the introduction of parallel instruments. Specifically, I define threshold levels beyond which redundancy of the EU ETS is to be expected. With knowledge of these threshold levels, policy makers are better informed and equipped to avoid a loss of control over European climate policy instruments.

7.2 Methodology

In this section, I describe the ETS model and the manner in which parallel instruments have been introduced into it. I build on the stochastic simulation model of the EU ETS described in Mulder and Jepma 2015. After a short summary of the original model, I offer a detailed description of the approach used to introduce Type 1 and Type 2 instruments into the model.

The stochastic EU ETS simulation model

The model of the EU ETS outlined in figure 7.1 simulates the fundamentals of the EU ETS, including annual abatement activity in various sectors and a forecast of the long-term carbon price (which I call the Fundamental Carbon Price Indicator). How the FCPI is calculated will be explained. The model runs from 2008 (the start of Phase II of the EU ETS) to 2030.

The supply of allowances mirrors the current regime for allocating annual allowances. The demand for allowances is equal to business-as-usual emissions reduced by abatement that is triggered by the EU ETS. (Abatement triggered by parallel instruments will be added to the model in a later subsection.) Realized demand levels since 2008 are exogenous input to the model. Starting in 2013, the business-as-usual growth in emission is sampled from a distribution via a Monte Carlo

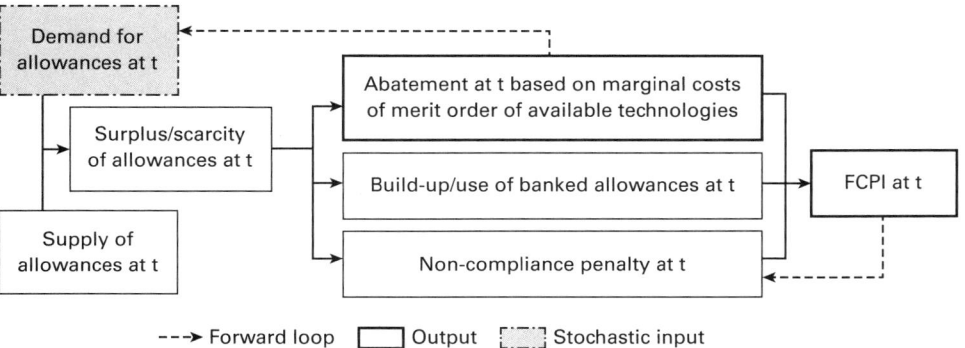

Figure 7.1
The original EU ETS model.

procedure. The distribution is based on historical growth rates of European industrial emissions between 1990 and 2008 (EEA 2009).

If the supply of allowances is larger than the demand, the surplus is banked for use in future years, because allowances do not expire. We assume that, if the demand for allowances exceeds the supply, firms have three options available to comply with ETS regulation: investing in carbon abatement, using previously banked allowances, and paying the non-compliance penalty.[3] Here, paying the non-compliance penalty is treated as the option of last resort: firms will pay the non-compliance penalty only if no other banked allowances or abatement opportunities are available. In case of non-compliance, the FCPI equals the non-compliance penalty of €100 per tonne of CO_2. Also, all remaining abatement potential and banked allowances will be utilized. In any other case, firms must choose between using banked allowances and investing in carbon abatement. Therefore, the extent to which firms choose either of the two options depends on their relative cost. An example is illustrated in figure 7.2, which shows the equilibrium between abatement and use of banked allowances in a random year t. In that year, the overall scarcity of allowances equals 88 $MtCO_2$ and is shown on the x axis. In equilibrium, the amount of abatement equals 51 $MtCO_2$, the use of banked allowances equals 37 $MtCO_2$, and the FCPI equals €43.

Note that we assume that firms make a minimum effort to comply with ETS regulation in order not to extract too many resources from their core business. This can be seen in figure 7.2, as more abatement potential is available at the same marginal cost yet not all potential is used.

The dotted curve representing demand for abatement is the merit-order abatement curve. All abatement opportunities are ordered according to their relative marginal cost in euros per metric tonne abated.

The curve representing supply of banked allowances rests on the following assumptions: Assume that the willingness to use banked allowances by firms at any time depends on firm-level carbon-price expectations and investment opportunities. Assume that firm-level carbon-price expectations and investment opportunities across Europe are heterogeneous, given that the population of firms under the EU ETS is highly diverse and operates under imperfect information. For example, some firms supply (hold on to) their banked allowances if the carbon price is €30 per tonne, given their relatively low (high) price expectation in the future or access to (lack of) alternative investment

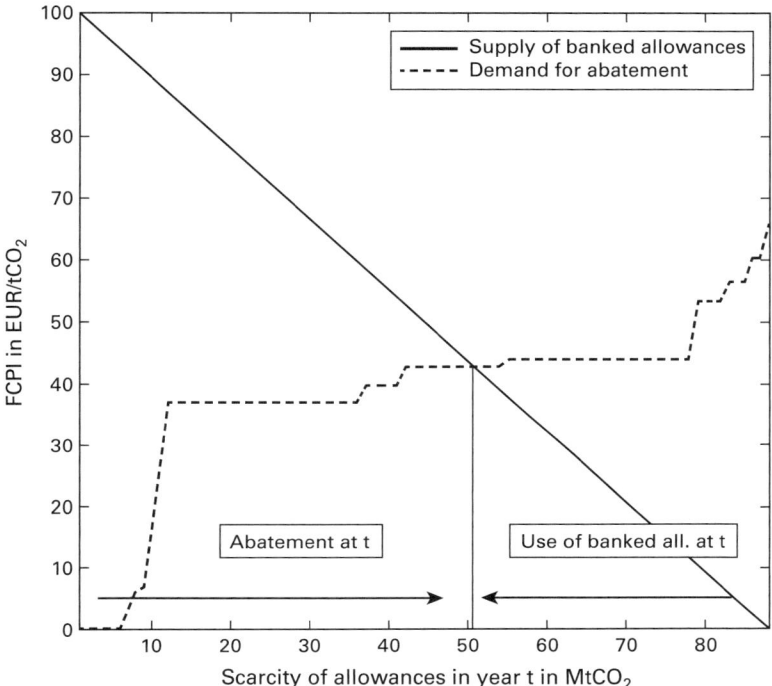

Figure 7.2
Abatement and use of banked allowances in randomly simulated year *t*.

opportunities with a higher expected return. Logically, the share of firms that are willing to supply their banked allowances increases as the carbon price goes up, as all firms are assumed to hold on to their banked allowances if the carbon price equals €0 per tonne while all firms are willing to supply their banked allowances if the market price equals the price ceiling (the non-compliance penalty). The curve representing supply of banked allowances is formed by assuming a linear relationship between these two extremes.

The equilibrium between the demand for abatement and the supply of banked allowances in a year determines the equilibrium FCPI in that year. The forecasted FCPI can be interpreted as a long-term forecast of the carbon price because the actual market price for carbon allowances is expected to converge to the FCPI. To understand why, consider a situation where the market price for allowances is considerably higher than the forecasted FCPI (e.g., due to speculation). This would trigger extra abatement activity and a reduction of demand below the equilibrium point, leading to a surplus of allowances. The surplus would put downward pressure on the carbon price, leading to convergence of the market price toward the FCPI.

Because allowance demand is uncertain, and because deployment of abatement technologies alters the shape of the merit order, a stochastic year-to-year carbon price pattern is formed. For an in-depth description of the model, see Mulder and Jepma 2015.

Introducing parallel instruments
The simulation window runs from 2008 to 2030. Within that window, we assume that parallel instruments are in place between 2013 and 2030.

Type 1 instruments
Type 1 parallel instruments are defined as instruments that lower the carbon intensity of production in ETS sectors. The common feature of these instruments is that they speed up investments in abatement technologies within ETS sectors. Introducing Type 1 instruments would affect the EU ETS, and thus would affect the ETS model, in two ways. First, allowance demand and scarcity would be reduced by the amount of abatement achieved with Type 1 instruments. Second, technologies from the merit order are used to achieve this reduction, so this abatement potential is no longer available in future periods, changing the shape of the merit order in figure 7.2.

It is not possible to individually model all the Type 1 instruments that are in operation today. Instead, I propose a method to estimate their aggregate effect on the functioning of the EU ETS. I assume that their collective impact ranges from 0 and 30 $MtCO_2$ of new abatement per year, as all relevant simulation results fall within that range.[4]

We run all possible scenarios within that range with increments of 1 $MtCO_2$. For each of these 31 scenarios the impact remains constant over time. Also, we assume in all scenarios that Type 1 instruments trigger deployment of all technologies in the merit order. For example, if Type 1 instruments are assumed to trigger a total of 10 $MtCO_2$ of new abatement per annum, the abatement potential of all technologies in the merit order is reduced proportionally until a reduction in emissions of 10 $MtCO_2$ is achieved. By reducing the abatement potential proportionately, I attempt to mirror the wide range of instruments that are in place today across industries, technologies, and abatement costs.[5]

Mathematically, the implications are rather straightforward. Allowance demand in now adjusted for the effect of Type 1 instruments:

$$AD_t^{new} = AD_t^{original} - T1_t. \tag{1}$$

Here $AD_t^{original}$ is allowance demand in year t as specified in the original model (see formula 2.2 in Mulder and Jepma 2015) and $T1_t$ is the reduction in demand through Type 1 instruments in year t in $MtCO_2$. The new allowance demand in year t, AD_t^{new}, is calculated by subtracting $T1_t$ from $AD_t^{original}$.

The effect of Type 1 instruments on the merit order is defined by the following two formulas:

$$\pi_t^i = \frac{TAC_t^{i,original}}{\sum_{k-1}^{n} TAC_t^{k,original}} \tag{2}$$

where π_t^i is a measure of the relative abundance of abatement technology i at t, $TAC_t^{i,original}$ is the total abatement capacity of technology i at t (in $MtCO_2$), as defined in the original model, and the denominator defines the cumulative capacity of all n abatement technologies that are available at t, and

$$TAC_t^{i,new} = TAC_t^{i,original} - \pi_t^i T1_t \tag{3}$$

where $TAC_t^{i,new}$ is the total remaining abatement capacity of technology i in year t; that is, the original abatement capacity of technology i is diminished by $\pi_t^i T1_t$, as that proportion is assumed to be deployed through Type 1 instruments.

Type 2 instruments

Type 2 instruments are defined as instruments that reduce the production levels in ETS sectors. Type 2 instruments are primarily found in non-ETS end-use sectors. Introducing Type 2 instruments would affect the EU ETS only by reducing allowance demand in ETS sectors, because firms in these sectors face lower demand for their end products.

Similar to Type 1 instruments, we assume that the collective impact of Type 2 instruments reduces the emission level by between 0 and 30 $MtCO_2$ per annum. We run all 31 scenarios within that range with increments of 1 $MtCO_2$, where the impact remains constant over time per scenario.

Mathematically, incorporating Type 2 instruments requires adding one more term to equation 1:

$$AD_t^{new} = AD_t^{original} - T1_t - T2_t \tag{4}$$

where $T2_t$ is the reduction in the demand for allowances through Type 2 instruments in year t in $MtCO_2$.

7.3 Results

As a reference, I first present results from the Neutral Scenario. In the Neutral Scenario, the ETS is the only instrument triggering carbon abatement activity; the annual impact of Type 1 and Type 2 instruments is thus 0 $MtCO_2$. In subsequent sections I examine the sensitivity of the carbon price and the sensitivity of the emission level and the probability that the EU ETS will become redundant.

In practice, both Type 1 and Type 2 instruments operate in a parallel fashion alongside the EU ETS. Therefore, every possible quantitative combination of Type 1 and Type 2 instruments within the specified range (0–30 $MtCO_2$/yr) is tested. As a result, I run a total of 961 (31 × 31) scenarios and present results from all scenarios in three-dimensional plots.

The Neutral Scenario

Figure 7.3 shows that the forecasted FCPI is equal to about €35 in 2008 but plummets in subsequent years because allowances are in oversupply in those years, primarily as a result of the financial and economic crisis. In absence of speculation, the price falls to zero. Around 2015, the mean FCPI quickly increases again, up to about €50 in 2025. In the last five years of the simulation, the FCPI decreases slightly as a result of technological learning, higher prices for fossil fuel, and greater availability of abatement technologies. Because allowance demand is uncertain, there is significant uncertainty around the forecasted equilibrium level of the FCPI, as reflected by the 80 percent confidence interval in figure 7.3. The effect of the financial and economic crisis is also clearly reflected in figure 7.4, which depicts the overall stock of banked allowances over time. Because no end date has been specified for the EU ETS, and allowances do not expire, a positive stock of allowances remains at the end of the simulation. The stock quickly builds up after 2008, and is then gradually reduced over time toward a mean of 2,000 $MtCO_2$ worth of allowances in 2030. In case of unusually strong economic growth between 2013 and 2030, a stock level below 1,500 $MtCO_2$ also remains a possibility.

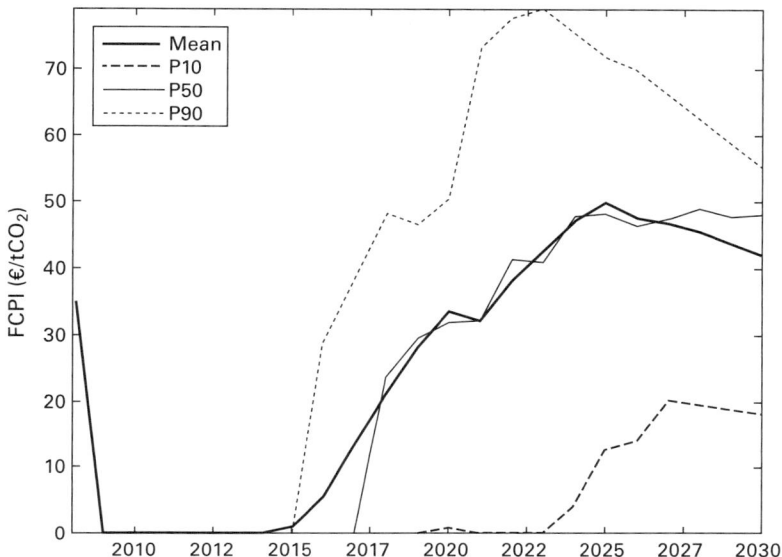

Figure 7.3
Fundamental EUA price—neutral scenario.

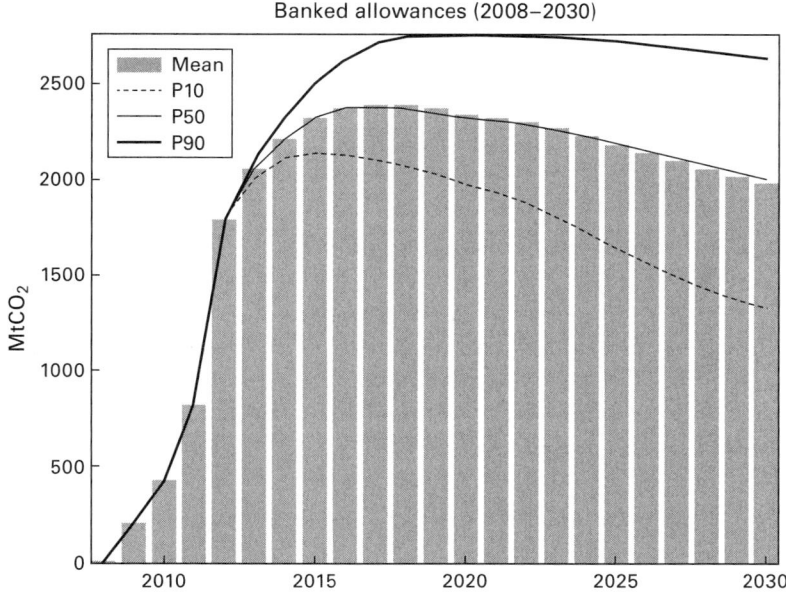

Figure 7.4
Stock of banked allowances—neutral scenario.

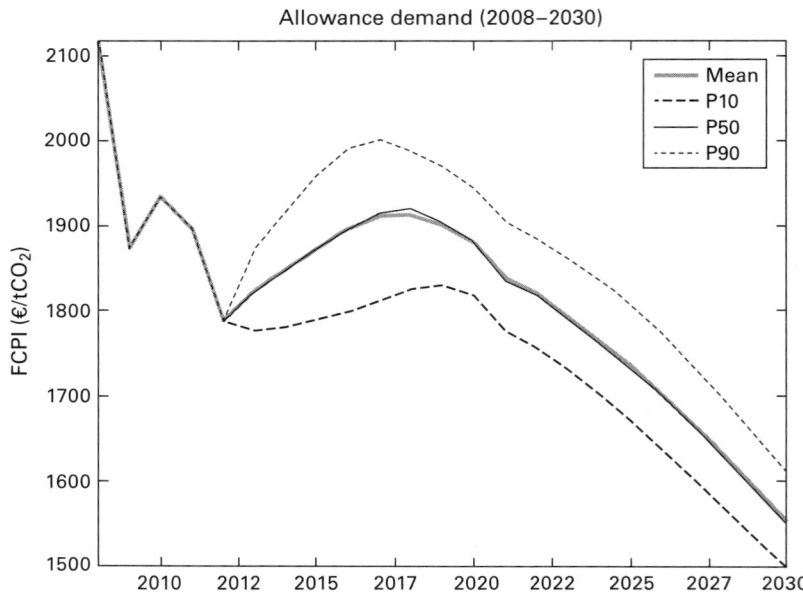

Figure 7.5
Emissions level under the EU ETS (in $MtCO_2/yr$)—neutral scenario.

The level of emissions is depicted in figure 7.5. Again, the effect of the economic crisis is clearly visible between 2008 and 2012. As a result, the emissions level has room to rebound until 2017. After 2017, the reduced supply of allowances forces firms to invest in carbon abatement, driving the emission level downward.

Effect of parallel instruments on carbon price

We present the mean and median forecasts of the FCPI that were obtained in all simulated scenarios to assess the impact of parallel instruments on the strength of the carbon price signal. Note that the depicted mean FCPI has limited value as a forecast of the actual carbon price because the market price is inherently uncertain and dependent on assumptions regarding the availability and marginal costs of specific technologies (Mulder and Jepma 2015).

The forecasted mean FCPIs in 2030 for all scenarios are shown in figure 7.6. The impact levels of Type 1 and Type 2 instruments (in $MtCO_2/yr$) are shown on the x and y axes respectively, with the mean FCPI shown on the z axis. The Neutral Scenario FCPI level in 2030 is €42. The FCPI is responsive to both the introduction of Type 1 and

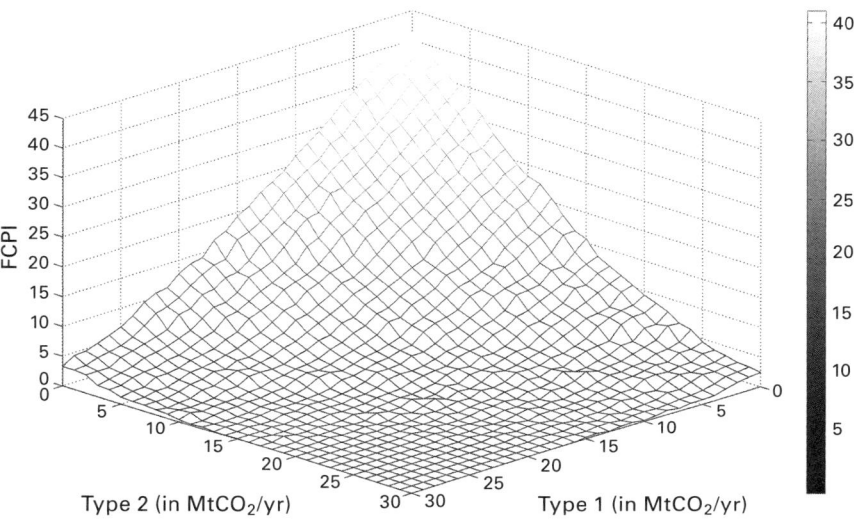

Figure 7.6
Effect of Type 1 and Type 2 instruments on the mean FCPI in 2030.

Type 2 instruments. As the combined impact of both types of instruments approaches 30 $MtCO_2$/yr, the mean FCPI approaches zero.

Whereas figure 7.6 depicts the case for 2030, figures 7.7 and 7.8 reveal that the FCPI is significantly lower in earlier years as well under the influence of parallel instruments. Two scenarios are depicted in each graph: the Neutral Scenario and a scenario with 20 $MtCO_2$ of abatement via Type 1 (figure 7.7) or Type 2 (figure 7.8) instruments.

The medians in figures 7.7 and 7.8 reveal that the probability distribution of the FCPI becomes positively skewed once parallel instruments are in effect. The medians effectively show that the carbon price is already likely to approach zero if the annual impact of parallel instruments is 20 $MtCO_2$/yr.

When figures 7.7 and 7.8 are compared with respect to the forecasted carbon price in the 20 $MtCO_2$/yr scenario, a lower mean FCPI can be seen in the latter figure, indicating that the carbon price is more sensitive to Type 2 instruments over time. The stronger carbon price sensitivity to Type 2 instruments can be explained by burden shifting between ETS and non-ETS sectors. To see why, consider that Type 2 instruments encourage non-ETS firms to pursue investments that reduce emissions in ETS sectors. Type 2 instruments thereby reduce the

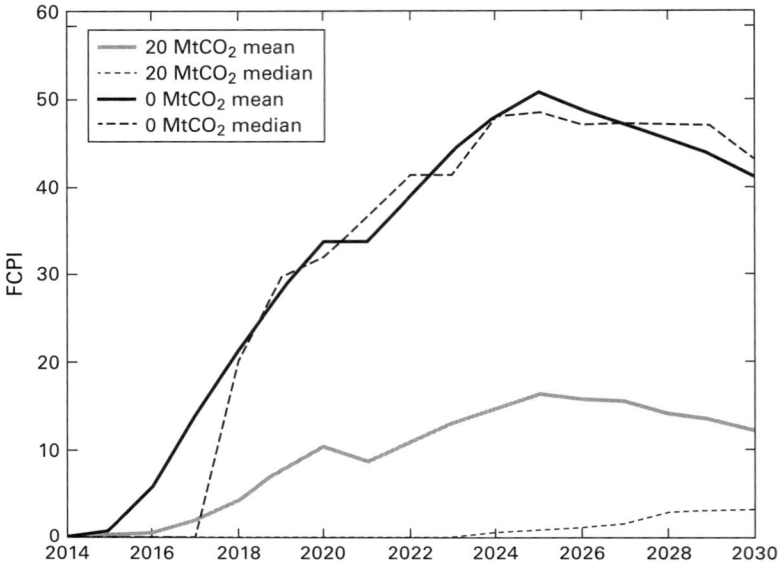

Figure 7.7
Mean and median FCPI assuming 0 and 20 $MtCO_2$ effect of Type 1 instruments. (Type 2 remains 0.)

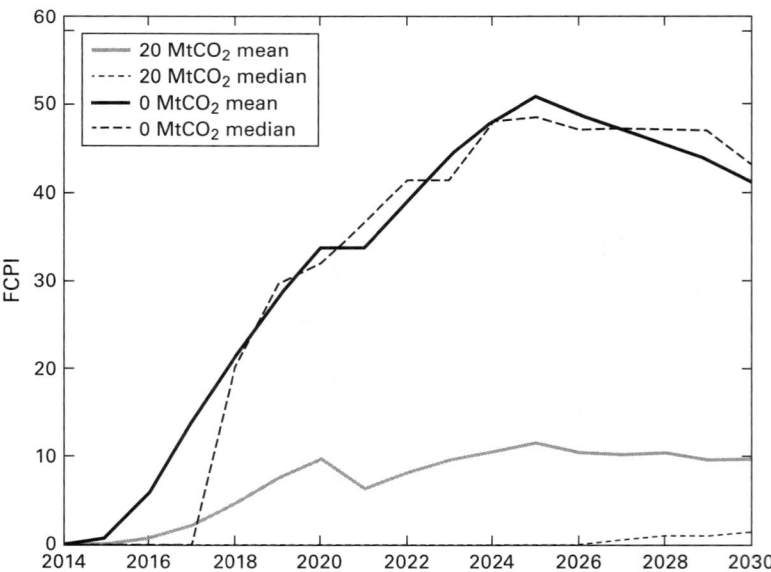

Figure 7.8
Mean and median FCPI assuming 0 and 20 $MtCO_2$ impact of Type 2 instruments. (Type 1 remains 0.)

need for ETS firms to invest in carbon abatement themselves to stay below the emission allowance cap. The reduced pressure on ETS firms is reflected by a lower carbon price, as ETS firms can comply with EU ETS regulation without having to invest in some of the more costly abatement technologies. Type 1 instruments, however, encourage ETS firms to invest in carbon abatement, just as the EU ETS does. Therefore, Type 1 instruments do not shift the abatement burden from ETS sectors to non-ETS sectors. Consequently the carbon price is higher in figure 7.7 than in figure 7.8.[6]

Effect of parallel instruments on achieved emission level

In this subsection, we present the mean forecasted emission levels within EU ETS sectors in 2030. Changes to this expected value following the introduction of parallel instruments reveal to what extent Type 1 and Type 2 instruments provide additional emission reduction alongside the EU ETS.

Figure 7.9 shows that a distinction can be made between scenarios with a relatively low impact level of parallel instruments and scenarios with a relatively high impact level of parallel instruments, although the exact threshold level that divides these two classes is hard to determine from figure 7.9 alone. Whereas the former class of scenarios does not

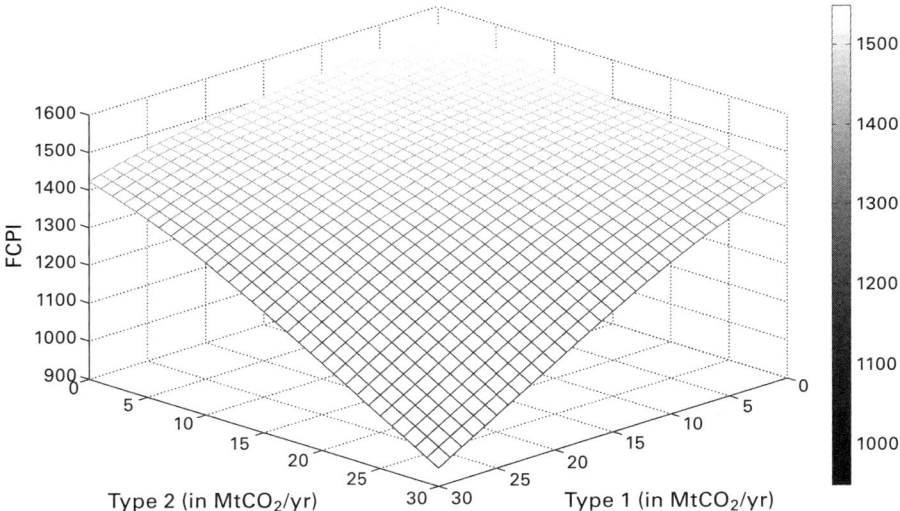

Figure 7.9
Emissions level within ETS sectors in 2030.

seem to have any significant influence on the emission level, the latter class of scenarios has a strong downward effect on the mean emission level. The two classes of scenarios will be described and discussed separately. Also, the threshold level will be determined and discussed in more detail.

Note that the emission level attained in the Neutral Scenario is sufficient to comply with EU ETS emission targets. Any emission reduction beyond that level, as a result of the introduction of parallel instruments, effectively means that governments are overshooting the EU emission target. Pursuing such an abatement strategy seems ill-advised because overshooting the target burdens European economies with unnecessarily high cost, and possible loss of competitiveness. If European policy makers are committed to achieving emission reduction beyond the current target level anyway, lowering the EU ETS allowance cap would be a more straightforward way to achieve that goal.

Scenarios with a relatively low impact of parallel instruments

If the gross impact via parallel instruments is relatively low, all abatement via parallel instruments is offset via the EU ETS. This occurs because the depreciated carbon price reduces abatement activity in ETS sectors to which the parallel instruments do not apply. For example, Type 1 instruments may speed up abatement activity in the power sector, which relieves the pressure on other ETS sectors (e.g., the steel or cement sector) to abate CO_2 and stay below the allowance cap. The net effect is that the emission level remains unchanged.

However, that offset of abatement activity via the EU ETS does not have to occur instantaneously. Parallel instruments typically have opposing effects on the emission level over time. As will be explained below, parallel instruments tend to speed up abatement activity in the short run but to slow it down in the long run. As a result of this intertemporal effect, the forecasted emission level in 2030 can be slightly below the Neutral Scenario level, even if the annual impact is very small. In practice, the small impact on the emission level can be explained by three factors.[7] First, construction lead times regarding abatement technology can delay the response in emissions output, despite an instantaneous response of the carbon price. Second, imperfect information on the emissions trading market could lead to a delayed downward response of the carbon price. Imperfect information relates to unawareness or the inability of market participants to

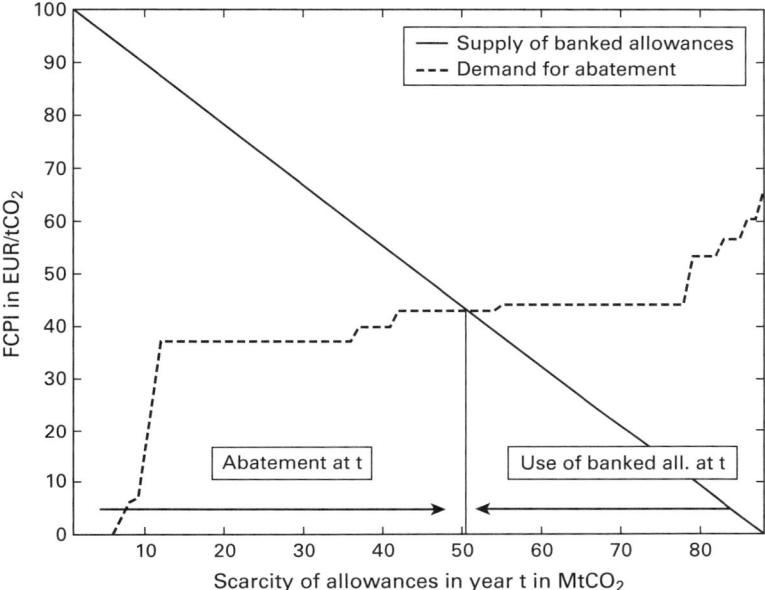

Figure 7.10
Abatement and use of banked allowances without parallel instruments in a random simulated year.

have full information regarding the future effect of instruments that are introduced by governing bodies across Europe. A third possible factor relates to the reduced option value of carbon allowances once parallel instruments depreciate the carbon price. To see why, consider the example illustrated in figures 7.10 and 7.11.

In figure 7.10, the overall scarcity (88 $MtCO_2$) is divided between abatement (51 $MtCO_2$) and use of banked allowances (37 $MtCO_2$), resulting in a carbon price of €43. Alternatively, if we assume that Type 1 and Type 2 instruments contribute 20 $MtCO_2$ to abatement efforts in that same year (as depicted in figure 7.11), industries under the EU ETS are not required to use as many banked allowances, or to invest in abatement, to comply with EU ETS regulation. The carbon price falls to €40 and abatement activity via the EU ETS is reduced to around 41 $MtCO_2$, while the use of banked allowances is reduced to 27 $MtCO_2$. If we compare the scenarios in figures 7.10 and 7.11, overall abatement efforts are higher in the latter scenario, as parallel instruments and the EU ETS trigger a total of 61 $MtCO_2$ (41 + 20 $MtCO_2$) whereas only 51 $MtCO_2$ is abated in the former scenario.

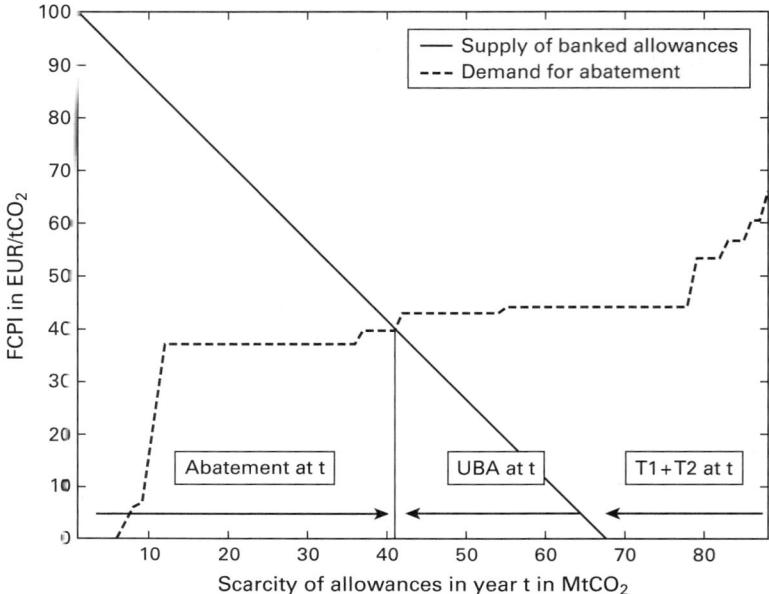

Figure 7.11
Effect of parallel instruments on abatement and use of banked allowances.

The boost in abatement activity in the latter scenario can be attributed to our assumption that firms have unchanged long-term carbon-price expectations.[8] Because firms do not lower their long-term carbon-price expectations, they have an incentive to hold on to their banked allowances as soon as the carbon price falls. In the end, they anticipate a higher option value in the future. Long-term carbon-price expectations might be unchanged because firms view parallel instruments as temporary (e.g., subsidies), assuming that the EU ETS remains the main instrument of European climate policy in the long term.

If we relax this assumption, the intertemporal effect on emissions may disappear or even change sign. An example of a scenario with relaxed assumptions is shown in figure 7.12. Now, the most optimistic firms under the EU ETS anticipate a long-term carbon price of €60 (instead of €100 in the previous scenarios). As a result, abatement via the EU ETS and parallel instruments totals 46 MtCO$_2$ (–5 MtCO$_2$ relative to the scenario in figure 7.10) and the use of banked allowances equals 41 MtCO$_2$ (+4 MtCO$_2$). This result indicates that, in addition to

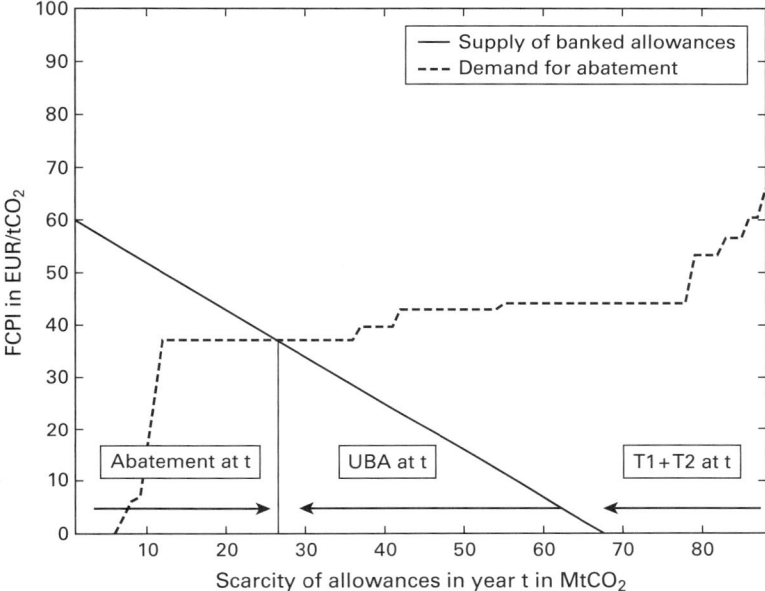

Figure 7.12
Parallel instruments, abatement, and use of banked allowances with lowered carbon-price expectations.

a temporary speedup, a temporary slowdown of abatement activity is also possible after the introduction of parallel instruments.

At an extreme, abatement activity via the EU ETS may come to a complete stop if ETS firms decide to dump their banked allowances. Such a scenario would become a possibility if ETS firms were to foresee the possibility of the EU ETS becoming redundant.

All in all, the scenarios above show that behavioral responses of ETS firms are crucial in determining whether a temporary speedup or slowdown of abatement activity may occur. In practice, it is not well understood how heterogeneous carbon-price expectations and allowance-banking strategies are affected by the introduction of parallel instruments. Nevertheless, the modeling exercise provides insight into possible explanations for a short-term speedup or slowdown in abatement activity. In our modeled scenarios we have assumed that long-term carbon-price expectations remain constant. Thus, we have implicitly assumed that firms commit to the EU ETS and anticipate that European policy makers will do the same.

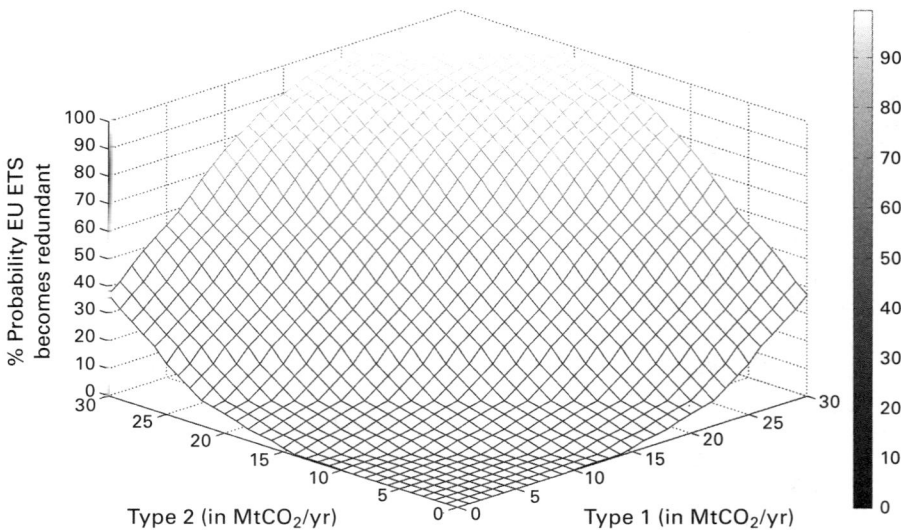

Figure 7.13
Probability that the EU ETS will become redundant.

Scenarios with a relatively high impact of parallel instruments

In figure 7.9, the mean emission level decreases steeply if the combined impact of parallel instruments is relatively high. This steep decrease can be explained by a higher probability that the EU ETS will become redundant. We define EU ETS redundancy as scenarios in which the EU ETS does not trigger any abatement activity until 2030. In those scenarios, the carbon price is forced down to zero.

The probability of EU ETS redundancy for each scenario is shown in figure 7.13. As that figure shows, as long as the annual impact remains below 20 $MtCO_2$, parallel instrument have too little impact to turn the EU ETS into a redundant scheme. If the annual impact of parallel instruments surpasses 20 $MtCO_2$, the EU ETS is in danger of becoming a redundant scheme; if the annual impact rises further, the probability rises steeply. Obviously, once the EU ETS has become a redundant scheme, abatement that is achieved via parallel instruments will no longer be offset via the EU ETS. This explains why, in figure 7.9, the mean forecasted emission level starts to decline if the annual impact of parallel instruments exceeds 20 $MtCO_2$.

Bear in mind that the numbers in figure 7.9 are mean forecasts. If economic growth were to remain lower than the mean until 2030, less abatement would be required to remain below the EU ETS allowance

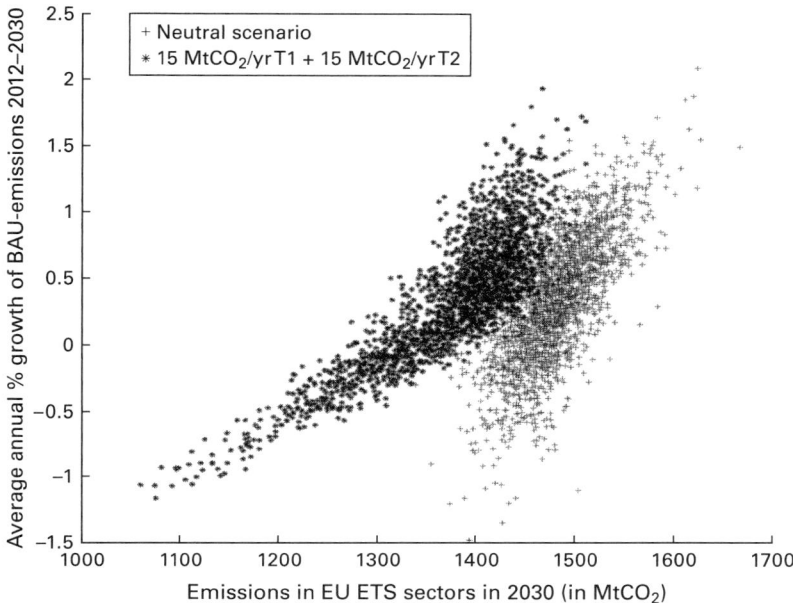

Figure 7.14
Effect of BAU emissions growth rate (largely driven by the economic growth rate) on the 2030 emission level.

cap. As a result, the EU ETS would also become redundant more easily. Figure 7.14 illustrates this phenomenon. It shows the forecasted emission level in 2030 as a function of the business-as-usual emissions growth rate between 2012 and 2030 for both the neutral scenario and for a scenario with a total impact by parallel instruments of 30 $MtCO_2$/yr.[9] Regarding the latter scenario, the tail at lower left shows that the impact of parallel instruments on the forecasted emission level increases significantly if economic growth is below the mean, which signals that the EU ETS has become redundant in these instances. At or above mean, economic growth rates the estimated emission level is also somewhat lower, but this can be attributed to a temporary boost in abatement activity and does not indicate that the EU ETS has become redundant.

The results plotted in figure 7.13 show that if the combined impact of parallel instruments surpasses 45 $MtCO_2$ per annum the EU ETS will undoubtedly become redundant. If the combined impact is in between 20 and 45 $MtCO_2$ per annum, the future of the EU ETS is uncertain and hinges on economic growth rates in Europe.

These threshold levels can be considered high estimates for two reasons. First, redundancy is defined rather strictly (zero abatement via the EU ETS). Even if a small amount of abatement is triggered by the EU ETS until 2030, reflected by a positive albeit low carbon price, policy makers are unlikely to remain as supportive of the EU ETS as they are today. The examples in figures 7.7 and 7.8 show that the carbon price is likely to be weak even if the combined impact of parallel instruments is just 20 MtCO$_2$. Second, in all scenarios, the long-term carbon-price expectations by firms are assumed to be unaffected by parallel instruments. Thus, we assume that firms commit to the EU ETS unconditionally, and do not anticipate its redundancy. Yet, in practice, firms could lose faith in the future of the EU ETS and dump their stock of banked allowances. All in all, if policy makers and firms take positions that further undermine the strength of the scheme, the actual threshold levels for redundancy may lie significantly below the levels estimated here.

7.4 Discussion

The results presented above reveal that the role of the EU ETS can be severely weakened if the aggregate impact of parallel instruments is 20 MtCO$_2$/yr. Note that 20 MtCO$_2$ is equal to a 1 percent reduction of annual emissions under the EU ETS. As a comparison, the EU ETS allowance cap is reduced by 1.74 percent per annum. Thus, even if the impact of parallel instruments were to be far below the intended reduction of the emissions cap, the EU ETS could be weakened to the extent that redundancy would become a real option, especially if some firms or policy makers were to lose their commitment to the scheme.

The factor that explains this phenomenon is the uncertainty of future economic growth and allowance demand: if economic growth stagnates, the chances of EU ETS redundancy increase sharply.

In previous studies of the redundancy of an ETS, stochastic demand patterns were not incorporated in the analysis. Also, the aggregate impact of multiple parallel instruments, as opposed to just one or two instruments, was largely overlooked. The combination of these two factors, however, seems to put the effectiveness of an emissions trading scheme in a less optimistic perspective.

The threshold levels that were estimated here offer European policy makers further tools for evaluating and controlling EU ETS performance. Rather than banning the use of parallel instruments, EU member states could introduce a cap on the use of parallel instruments based on these threshold levels. That is, a cap (in $MtCO_2$ of abatement achieved per annum) could be set on the maximum allowed use of parallel instruments in each member state. In that manner, the total impact of parallel instruments could remain well below the threshold level. Although its implementation is unlikely to be an easy exercise, a "cap on parallel instruments" would have two advantages. First, it would ensure that the economic and competitive advantages of cooperation via the EU ETS would be reaped, as possible unintentional redundancy of the EU ETS would be avoided. Second, the cap would stimulate national governments to introduce only those parallel instruments that would offer the greatest local benefits. In that manner, local governments would be forced to allocate public resources in a more cost-efficient manner, in the process making them aware of adverse interactions between the EU ETS and parallel instruments.

It would be valuable to examine whether a cap on parallel instruments is compatible within a policy setting with multiple energy and climate targets. More specifically, if the use of parallel instruments were to be restricted, stringent renewables and energy efficiency targets might become unachievable. If that were to happen, it might imply that policy makers will first have to lower the EU ETS allowance cap to allow for a higher cap on the use of parallel instruments. Alternatively, policy makers could set less ambitious renewables and energy efficiency targets. In either case, a cap on parallel instruments would give policy makers an incentive to design a more coherent policy setting in which adverse interactions would be reduced to a minimum and public resources would be used in a cost-efficient manner. In the process, a loss of control would be avoided.

As long as a cap on the use of parallel instruments is not in place, the results suggest, policy makers should be careful with the introduction of parallel instruments alongside the EU ETS if the scheme is intended to play an important role. Aggressively passing and implementing non-EU ETS instruments while carbon prices are already low could block an upward trend in the carbon price and ultimately lead to the redundancy of the EU ETS.

It should be noted that the aggregate impact of parallel instruments has been of a varying theoretical magnitude in this study. In view of the possibly strong implications of the results for environmental policy design, the results seem to justify more empirical research.

Conclusion

The results show that if the aggregate abatement impact of parallel instruments alongside the EU ETS is below approximately 20 $MtCO_2$/yr the forecasted mean emission level remains unaffected. This occurs because all abatement via parallel instruments is offset via the EU ETS.

The forecasted emission level does decrease significantly if the aggregate impact of parallel instruments is greater than 20 $MtCO_2$/yr. This can be explained by a greater probability that the EU ETS will become redundant. The EU ETS is assumed to be redundant if the scheme fails to trigger any abatement between 2013 and 2030. If the combined impact of parallel instruments is in between 20 and 45 $MtCO_2$ per annum, the future of the EU ETS is uncertain, and hinges on the overall economic growth rates in Europe. The lower the average rate of economic growth, the greater the likelihood that the EU ETS becomes redundant. Redundancy of the EU ETS is certain if the combined impact of parallel instruments surpasses 45 $MtCO_2$ per annum. We have multiple reasons to believe that these estimates are conservative, and that the actual threshold levels could be significantly below the reported figures. When differentiating types of parallel instruments, the results show that Type 2 instruments lead to a stronger depreciation of the EU ETS carbon price than Type 1 instruments. This can be explained by the fact that the former type of instruments lead to burden shifting between sectors: investments by end-use sectors effectively reduce the need for ETS sectors to invest in carbon abatement. If policy makers prioritize strong EU ETS performance, the results suggest, they should refrain from introducing parallel instruments in both ETS and non-ETS sectors if the carbon price is already weak.

Acknowledgments

I would like to thank Catrinus Jepma, Erik Dietzenbacher, Steven Brakman and Christian Bos for their valuable input during the writing process.

The work reported in this chapter been conducted in the framework of the CATO-2 project. CATO-2 is the Dutch national research program on CO_2 Capture and Storage. The program is financially supported by the Dutch Ministry of Economic Affairs and by the parties in the CATO-2 consortium.

Notes

1. The renewables quota is enforced through a "green certificate" scheme. The threshold levels depends on the stringency of the emissions trading scheme.

2. Hindsberger et al. (2003) examine a situation in which an international emissions trading scheme partly overlaps geographically with an international scheme of "tradable green certificates" to stimulate the deployment of renewable energy in the Baltic Sea region. They find that the carbon price approaches zero if the renewable energy target is set sufficiently high.

3. Carbon leakage to non-EU member countries is not included in the model.

4. Emissions in ETS sectors were 1,898 $MtCO_2$ in 2012. Therefore, the tested range (0–30 $MtCO_2$/yr) is equivalent to abatement between 0 and 1.6 percent per annum.

5. For example, Europe is attempting to stimulate a relatively expensive technology such as CO_2 capture and storage through subsidies while efficiency improvements (which typically have a low marginal cost) also receive support via a wide range of European programs in line with its 20–20–20 targets (EP 2010).

6. Although Type 1 instruments do not shift the *abatement* burden from ETS firms to non-ETS firms, Type 1 instruments may shift part of the *financial* burden to the government if a Type 1 instrument consists out of public financial assistance (e.g., subsidies). However, in this study I limit my analysis to the overall effect of parallel instruments on the performance of the EU ETS (notably the carbon price); the financing structure of individual investments is outside the scope of this research.

7. The first two factors are not modeled here; the third factor is accounted for in the simulation and will be discussed in more detail.

8. The "supply of banked allowances" curve still intersects the y axis at €100 per tonne.

9. The business-as-usual emissions growth rate is to a large extent driven by the economic growth rate, which we have not modeled explicitly. However, low values of the business-as-usual emissions growth rate are observationally equivalent to scenarios with a below-average economics growth rate.

References

Abrell, J., and H. Weigt. 2008. The interaction of emissions trading and renewable energy promotion. Working paper WP-EGW-05, Technische Universität Dresden.

Acocella, N., G. Di Bartolomeo, and A. Hughes Hallett. 2012. *The Theory of Economic Policy in a Strategic Context*. Cambridge University Press.

Amundsen, E. S., and J. B. Mortensen. 2001. The Danish Green Certificate System: Some simple analytical results. *Energy Economics* 23: 489–509.

Anandarajah, G., and N. Strachan. 2010. Interactions and implications of renewable and climate change policy on UK energy scenarios. *Energy Policy* 38: 6724–6735.

Bennear, L. S., and R. Stavins. 2007. Second-best theory and the use of multiple policy instruments. *Environmental and Resource Economics* 37: 111–129.

Böhringer, C., H. Koschel, and U. Moslener. 2008. Efficiency losses from overlapping regulation of EU carbon emissions. *Journal of Regulatory Economics* 33: 299–317.

Boots, M. 2003. Green certificates and carbon trading in the Netherlands. *Energy Policy* 31: 43–50.

Bye, T., and A. Bruvoll. 2008. Multiple instruments to change energy behaviour: The emperor's new clothes? *Energy Efficiency* 1: 373–386.

Conrad, K., and R. E. Kohn. 1996. The US market for SO_2 permits: Policy implications of the low price and trading volume. *Energy Policy* 24: 1051–1059.

De Jonghe, C., E. Delarue, R. Belmans, and W. D'haeseleer. 2009. Interactions between measures for the support of electricity from renewable energy sources and CO_2 mitigation. *Energy Policy* 37: 4743–4752.

Del Rio, P. 2007. The interaction between emissions trading and renewable electricity support schemes. An overview of the literature. *Mitigation and Adaptation Strategies for Global Change* 12: 1363–1390.

Del Rio, P. 2009. Interactions between climate and energy policies: The case of Spain. *Climate Policy* 9: 119–138.

Di Bartolomeo, G., A. Hughes Hallett, and N. Acocella. 2011. Tinbergen controllability and n-player LQ-games. *Economics Letters* 113: 32–34.

EEA. 2009. Annual European Union greenhouse gas inventory 1990–2009 and inventory report 2011. European Environment Agency, Technical Report No.2.

EEA. 2011. Resource efficiency in Europe Policies and approaches in 31 EEA member and cooperating countries. European Environmental Agency Report, No 5/2011.

European Parliament. 2010. Overview of Energy Efficiency measures of European Industry. Directorate-General for Internal Policies, IP/A/ITRE/NT/2010-08.

Hindsberger, M., M. H. Nybroe, H. F. Ravn, and R. Schmidt. 2003. Co-existence of electricity, TEP, and TGC markets in the Baltic Sea Region. *Energy Policy* 31: 85–96.

Jensen, S. G., and K. Skytte. 2003. Simultaneous attainment of energy goals by means of green certificates and emission permits. *Energy Policy* 31: 63–71.

Kautto, N., A. Arasto, J. Sijm, and P. Reck. 2012. Interaction of the EU ETS and national climate policy instruments—Impact on biomass use. *Biomass and Bioenergy* 38: 117–127.

Lundberg, H., V. Corless, D. Havlikova, B. Laird, and L. Rygnestad. 2012. *Renewable Energy Policies in Europe A Mapping of Existing and Planned Support Schemes for Renewable Energy Development in 31 European Countries*. Bellona Foundation.

Morris, J., J. Reilly, and S. Paltsev. 2010. Combining a renewable portfolio standard with a cap-and-trade policy: A general equilibrium analysis. Report 187, MIT Joint Program on the Science and Policy of Global Change.

Morthorst, P. E. 2001. Interactions of a tradable green certificate market with a tradable permits market. *Energy Policy* 29: 345–353.

Morthorst, P. E. 2003. National environmental targets and international emission reduction instruments. *Energy Policy* 31: 73–83.

Majone, G. 1989. *Evidence, Argument, and Persuasion in the Policy Process.* Yale University Press.

Mulder, A. J., and C. J. Jepma. 2015. Stochastic simulation of CO_2 emissions trading in Europe: Will the EU ETS drive investments in CCS? In A. J. Mulder, CO_2 Trading in the EU: Models and Policy Applications, doctoral dissertation, University of Groningen.

Oikonomou, V., and C. J. Jepma. 2008. A framework on interactions of climate and energy policy instruments. *Mitigation and Adaptation Strategies for Global Change* 13: 131–156.

Pizer, W. A. 2002. Combining price and quantity controls to mitigate global climate change. *Journal of Public Economics* 85: 409–434.

Rathmann, M. 2007. So support systems for RES-E reduce EU-ETS-driven electricity prices? *Energy Policy* 35: 342–349.

Interact. 2003. Interaction in EU climate policy. Science Policy Research Unit, University of Sussex.

Sorrell, S., and J. Sijm. 2003. Carbon trading in the policy mix. *Oxford Review of Economic Policy* 19: 420–437.

Theil, H. 1954. Econometric models and welfare maximization. *Weltwirtschaftliches Archiv* 72: 60–83.

Theil, H. 1956. On the theory of economic policy. *American Economic Review* 46: 360–366.

Theil, H. 1964. *Optimal Decision Rules for Government and Industry.* North-Holland.

Tinbergen, J. 1952. *On the Theory of Economic Policy.* North-Holland.

Tinbergen, J. 1956. *Economic Policies, Principles and Design.* North-Holland.

Wildavsky, A. 1979. *The Art and Craft of Policy Analysis.* Macmillan.

IV Firm Behavior

8 The Relationship between Spot and Futures CO$_2$ Emission Allowance Prices in the EU ETS

Stefan Trück, Wolfgang Härdle, and Rafał Weron

In January 2005 the advent of the EU-wide emissions trading scheme introduced emission allowances as a new class of financial assets. Since environmental policy has historically been a command-and-control-type regulation strictly requiring companies to comply with emission standards, the trading scheme indicates a shift in paradigms. The market not only requires regulated emitters to run an adequate risk management, it also provides new business development opportunities for market intermediaries and service providers such as brokers or marketeers.

In the EU ETS, described in detail in chapter 1, an installation's failure to submit a sufficient amount of allowances resulted in sanction payments of €40 per missing ton of CO$_2$ allowances during the pilot period and €100 in the second and third trading periods. Hence, the carbon emission market forces companies to hold an adequate number of allowances according to their carbon dioxide output. As a consequence, participating companies face several risks specific to emissions trading. In particular, price risk (of fluctuating allowance prices) and volume risk (because of unexpected fluctuations in energy demand, the emitters do not know ex-ante their exact demand for EUAs) have to be considered. Naturally, market generic risks—counterparty, operational, reputational, and so on—are also present. Participating companies may need to develop adequate risk management strategies as well as reliable models for the demand and for CO$_2$ allowance prices to reduce the risk of facing substantial sanction payments or possible high prices for purchasing additional CO$_2$ allowances. For a thorough discussion of this issue, see Bokenkamp et al. 2005.

Clearly, in addition to monitoring the spot market, market participants seeking to hedge against these risks will take great interest in derivative instruments such as options and futures contracts for carbon emission allowances. Since these participants must decide when to sell surplus or buy additionally required permits, the relationship between carbon spot and futures prices is of particular interest. For other commodities, including oil and agricultural products, the connection between spot and futures prices and the convenience yield has been investigated more thoroughly. For pricing contingent claims in commodity markets, Gibson and Schwartz (1990) present a two-factor model using the spot price and the instantaneous convenience yield as factors. With respect to the relationship between spot and futures prices, the literature finds some evidence on expected spot prices often exceeding the futures price of such assets (Bodie and Rosansky 1980; Chang 1985; Pindyck 2001). This situation, called *normal backwardation*, was initially suggested by Keynes (1930). Wei and Zhu (2006) find economically significant convenience yields and risk premiums in the US natural gas market. However, for electricity prices there is also some evidence that futures prices may exceed expected spot prices (Bierbrauer et al. 2007; Botterud et al. 2010; Longstaff and Wang 2004; Weron 2008; Weron and Zator 2014), a situation called *contango*. Owing to the peculiarity of the market for CO_2 emission allowances and the ambiguous results in different commodity markets, it seems worthwhile to compare the behavior of EUA spot and futures prices.

Since the official start of spot and futures trading in 2005, there have been many studies of the price behavior of CO_2 spot or futures contracts. Less attention has been directed to the relationship between the two markets.

Paolella and Taschini (2008), Seifert et al. (2008), and Benz and Trück (2008) were among the first to provide an econometric analysis of the behavior of allowance prices and investigate different models for the dynamics of short-term spot prices. Paolella and Taschini (2008) conduct an econometric analysis of emission allowance spot market returns and suggest that the conditional dynamics of allowance returns can be approximated by a GARCH-type structure. Seifert et al. (2008) develop a theoretical stochastic equilibrium model in order to incorporate stylized facts of the European carbon market. Their findings suggest that discounted CO_2 prices should possess the martingale property and that carbon prices exhibit a time- and price-dependent volatility structure. Benz and Trück (2008) apply autoregressive GARCH and regime-switching models to describe the dynamics of Phase I emission allowance spot prices. Their results support the adequacy of the models considered, capturing skewness, excess kurtosis, and different phases of volatility behavior in the returns.

Another stream of literature is more concerned with the drivers of allowance prices. Chevallier (2009a) examines the empirical relationship between the returns on carbon futures and changes in macroeconomic conditions and documents that carbon futures returns may be weakly forecast on the basis of two variables from the stock and bond markets, i.e., equity dividend yields and the "junk bond" premium. Hintermann (2010) examines price drivers of EUAs during the first phase of the EU ETS, in particular the extent to which variation in prices can be explained by marginal abatement costs. Bredin and Muckley (2011) examine the extent to which fundamental factors, including economic growth, energy prices, and weather conditions, determined EUA futures prices during the period 2005–2009. Chevallier (2011) suggests that yearly compliance events and the increasing uncertainties in post-Kyoto international agreements may explain the instability in the volatility of carbon prices. Gronwald et al. (2011) investigate the dependence structure between EUA futures returns and those of other commodities, equity indices, and energy indices and find significant positive correlations between the markets considered. Conrad et al. (2012) model the adjustment process of EUA prices to the releases of announcements at high frequency. They find that decisions of the European Commission on National Allocation Plans have a strong and immediate effect on EUA prices and that EUA prices increase in response to better-than-expected news about future economic development.

A number of studies provide insights on the pricing of derivative instruments for emission allowances. Daskalakis et al. (2009) develop a framework for the pricing and hedging of intra-phase and inter-phase derivatives contracts and find some evidence that market participants adopt standard no-arbitrage pricing. Carmona and Hinz (2011) develop a risk-neutral reduced-form model for allowance futures prices and show how to price European call options written on these contracts. Chesney and Taschini (2012) use dynamic optimization models to endogenously generate the price dynamics of emission permits under asymmetric information, allowing inter-temporal banking and borrowing. The model is solved numerically and a closed-form pricing formula for European-style options is derived. In an empirical study, Isenegger et al. (2013) evaluate different models for the pricing of exotic option contracts based on observed carbon spot and futures prices. Approaches considered in their study include a standard GARCH model, a threshold GARCH model, and a mean-reverting jump diffusion model. On the basis of observed market prices, Isenegger et al. suggest that the mean-reverting jump diffusion model provides the best fit to the option instruments analyzed.

There also have been several studies of price discovery in CO_2 spot and futures markets. Uhrig-Homburg and Wagner (2009), analyzing data from the pilot trading period, find that futures contracts expiring in December 2006 and 2007 led the price discovery process of CO_2 emission certificates. On the other hand, Milunovich and Joyeux (2010), also examining data from Phase I, suggest that CO_2 spot and futures markets share information efficiently and contribute to price discovery jointly. These results are confirmed by Niblock and Harrison (2012), who also find that spot and forward prices both contribute to price discovery in carbon markets during the pilot trading period. Gorenflo (2013) conducts an impulse response analysis and finds that the futures market has a leadership position against the spot market and contributes the most to price discovery. Benz and Hengelbrock (2008) analyze price discovery between the ECX and Nord Pool exchanges using intraday data. They suggest that for EUA futures contracts with maturity in December 2005 and 2006 both exchanges contributed to price discovery. On the other hand, for contracts maturing in December 2007 and December 2008, they find that the more liquid market ECX is leading the less liquid Nord Pool market.

Most relevant to the topic of this chapter is previous empirical research on the relationship among spot and futures prices, convenience yields, and deviations from the cost-of-carry relationship in emissions allowance markets. Milunovich and Joyeux (2010) examine the issues of market efficiency in the EU carbon futures market during the pilot trading period. The authors find that none of the carbon futures contracts examined are priced according to a cost-of-carry model. However, futures contracts referring to the pilot trading period form a stable long-run relationship with the spot price and can be considered risk-mitigation instruments. Uhrig-Homburg and Wagner (2009), also examining EUA prices during the pilot trading period, find contradictory results: examining the relationship between EU carbon spot and futures markets during Phase I, they suggest that the cost-of-carry model largely holds after an initial period of rather noisy pricing. They report that, although the convenience yield is not consistent over time and although there may be temporary deviations from the cost-of-carry linkage, the deviations generally vanish after a few days. Unfortunately, the results of these two studies are limited to the first trading period, in which banking of allowances from the pilot period to the Kyoto commitment period was not allowed. Therefore, results on the cost-of-carry relationship between spot and futures contracts might be questionable, in particular when looking at inter-period relationships. Chevallier (2009b) investigates the modeling of the convenience yield in the EU ETS for Phase II, using daily and intra-daily measures of volatility, finds a non-linear relation between spot and futures prices, and suggests that the dynamics of the observed convenience yield can be best described by a simple autoregressive process. Madaleno and Pinho (2011) examine EUA spot and futures prices from an ex-post perspective, also for the first Kyoto commitment period, and find evidence of a significant negative risk premium (i.e., a positive forward premium) in the market. They also find a positive relationship between risk premiums and time to maturity of the futures contracts. More recently, Gorenflo (2013) suggests that the cost-of-carry hypothesis between spot and futures prices holds for the trial period whereas for the Kyoto commitment period there are deviations from the cost-of-carry relationship. Chang et al. (2013), using the cost-of-carry model, examine the properties of convenience yields for emissions allowances futures contracts with maturities from December 2010 to December 2014. Chang et al. suggest that convenience yields for CO$_2$ emissions allowances exhibit a time-varying trend and are mean-reverting, and that the standard deviation in the convenience yield declines with an increase in time to maturity.

This chapter, to the best of our knowledge, is the first study of the relationship between carbon emission spot and futures prices to consider both pilot trading from 2005 to 2007 and the entire first Kyoto commitment period (2008–2012). Our goal is to provide a thorough analysis of the relationship between spot and futures prices in the EU ETS, also in comparison to other commodity markets. We investigate correlations between spot and futures contracts, deviations from the cost-of-carry relationship, convenience yields, and the volatility term structure for these two periods. We relate our results to such general concepts of commodity markets as backwardation and contango markets. We examine how changes in regulations pertaining to the banking of emission allowance contracts between Phase I and Phase II have affected the relationship between spot and futures prices. By investigating these issues, we also provide insights into participants' evaluation of risks in the market, their reaction to price shocks, and their assumptions about the future supply of and demand for permits during the second Kyoto commitment period.

8.1 Commodity spot and futures markets

An appropriate approach to specifying EUAs might be to consider them as a factor of production. (See, e.g., Fichtner 2004 or Benz and Trück 2006.) Similar to other commodities, they can be "exhausted" for the production of CO_2 and removed from the market after their redemption. Since a competitive commodity market is subject to stochastic fluctuations in both production and consumption, market participants will generally hold inventories. For emission allowances, producers may hold such inventories to reduce the costs of adjusting production over time or to avoid stockouts. In contrast with other factors of production, the amount of allowances is not required to match the actual production figure of the preceding calendar year until April 30 of the next year. However, examining appropriate financial models for CO_2 emission allowances, the obvious parallels to a factor of production motivate the idea to adopt approaches from commodity markets rather than using typical financial models for asset pricing. Hence, in this section we will briefly review some features of commodity markets, focusing on the relationship between the spot and futures markets.

Backwardation and contango

The futures market is said to exhibit *backwardation* when the futures price $F_{t,T}$ is less than or equal to the current spot price S_t; it exhibits *normal backwardation* when the futures price is less than or equal to the expected spot price $E_t(S_T)$ at time T. On the other hand, the term *normal contango* is used to describe the opposite situation, in which the futures price $F_{t,T}$ exceeds the (expected) spot price at time T; see, for example, Hull 2005 or Pindyck 2001. Table 8.1 summarizes the four situations.

Contango and backwardation markets can be motivated by market participants' need for hedging. In normal backwardation the producers

Table 8.1
Description of market situation based on the relationship between (expected) spot and futures price.

Market situation	Relation between (expected) spot and futures price
Backwardation	$F_{t,T} \leq S_t$
Normal Backwardation	$F_{t,T} \leq E_t(S_T)$
Contango	$F_{t,T} > S_t$
Normal contango	$F_{t,T} > E_t(S_T)$

are buying insurance against falling prices, whereas in contango consumers typically buy insurance against rising prices. The theory postulates that commodity futures markets usually exhibit backwardation and tend to rise over the life of a futures contract. Initially suggested by Keynes (1930) and Hicks (1946), the idea of backwardation assumes that hedgers tend to hold short positions as insurance against their cash position and must pay speculators a premium to hold long positions in order to offset their risk. Thus, observed futures prices $F_{t,T}$ with delivery at time T are often below the expected spot price $E_t(S_T)$. Because the risk is transferred to the long position in the futures contract, normal backwardation is equivalent to a positive risk premium; likewise, normal contango is equivalent to a negative risk premium. The risk premium is defined formally as the reward for holding a risky investment rather than a risk-free one. In other words, the risk premium is the difference between the expected spot price (the best estimate of the going rate of the asset at some specific time in the future) and the forward price (the actual price a trader is prepared to pay today for delivery of the asset in the future) (Botterud et al. 2010; Diko et al. 2006; Pindyck 2001; Weron and Zator 2014). In the literature on financial mathematics, yet another notion is used. The *market price of risk* (often denoted by λ) is defined as the difference between the drift in the original "risky" probability measure P and the drift in the "risk-neutral" measure P^λ in the stochastic differential equation governing the price dynamics (Weron 2006). The spot price forecast $E_t(S_T)$ is the expected value of the spot price at some future date with respect to P, and the forward price $F_{t,T}$ is the expected value of the spot price with respect to P^λ. If λ is positive then the risk premium is also positive, and vice versa.

The term structure of a commodity's forward price volatility is also interesting. Samuelson (1965) found a typically declining term structure in the volatility of futures prices as maturity increases. This behavior is referred to as the *Samuelson effect* or as the *time-to-maturity effect*. The behavior is generally explained by the fact that the opinion of investors of a distant future environment, including the evaluation of distant futures prices, is subject to minor changes only in the near future. Thus, it is assumed that only a few of the parameters affecting the final level of the prices will change today. Hence, only minor effects can be expected for futures with long maturities. However, as the maturity date approaches, investors are clearly more sensitive to information that influences the level of the futures price at maturity.

Relating spot and futures prices

Approaches to the valuation of forward and futures contracts can be conceptually divided into two groups (Fama and French 1987). The first group suggests a risk premium to derive a model for the relationship between short-term and long-term prices. The second group is closely linked to the cost and convenience of holding inventories. In this sub-section we follow the second approach and briefly illustrate the derivation of the convenience yield.

The convenience yield is usually derived within a no-arbitrage or cost-of-carry model that is based on considerations on a hedging strategy consisting of holding the underlying asset of the futures contract until maturity. The long position in the underlying asset is then funded by a short position in the money-market account. Risk drivers determining the futures price in this case include the cost of storage for forward on commodities, the cost of delivery, and the interest-rate risk. Differences between current spot prices and futures prices are explained by interest forgone in storing a commodity, warehousing costs, and the so-called convenience yield on inventory. By assuming no possibilities for arbitrage between the spot and futures market, a formula for the convenience yield can be derived (Geman 2005; Pindyck 2001).

Assume that we hold one unit of emission rights at time t and that the current spot price is S_t. Obviously there is no physical storage cost for holding an emission right. Hence, assuming the existence of a convenience yield, holding the emission right until T will pay us the stochastic return

$$S_T - S_t + \gamma_{(T-t)}. \tag{1}$$

Here $\gamma_{(T-t)}$ denotes the convenience yield for holding the emission right from t until T. Assume that at the same time we also short a futures contract written on the emission right with delivery in T. The return of this futures contract equals $F_{t,T} - S_T$. Note that there is no risk involved in the transactions and the total return is non-stochastic and should equal the return on a risk-free investment with value S_t for the period from T-t:

$$S_T - S_t + \gamma_{(T-t)} + F_{t,T} - S_T = (e^{r(T-t)} - 1)S_t. \tag{2}$$

Solving for $\gamma_{(T-t)}$, we get the following equation for the capitalized flow of marginal convenience yield (Pindyck 2001):

$$\gamma_{(T-t)} = S_t e^{r(T-t)} - F_{t,T}. \tag{3}$$

The convenience yield obtained from holding a commodity can be regarded as similar to the dividend obtained from holding a company's stock. It represents the privilege of holding a unit of inventory, for instance, to be able to meet unexpected demand. According to Pindyck (2001), the spot price of a commodity can be thought of as similar to the price of a stock: as the price of a stock can be regarded as the present value of the expected future flow of dividends, the price of a commodity is the present value of the expected future flow of convenience yields. Alternatively, one could argue that the convenience yield is the *residual* needed to align cost-of-carry commodity futures prices with observed market prices. The cost-of-carry model describes an arbitrage relationship between the futures price, the spot price, and the cost of carrying the asset, so that with zero cost of storage the cost-of-carry relationship can simply be expressed as

$$F_{t,T} = S_t e^{r(T-t)} - \varepsilon_{t,T}. \tag{4}$$

Clearly, by comparing equation 3 and equation 4 one can see that the convenience yield equals the residual $\varepsilon_{t,T}$ in the cost-of-carry model.

8.2 Empirical results

The data
For our analysis we use available market quotes from the European Energy Exchange (EEX) and the European Climate Exchange (ECX) for the pilot trading period. Our price data for the Kyoto commitment period are from PointCarbon, a major supplier of data for the global gas, power, and carbon markets. More precisely, data on Phase I spot prices are obtained from EEX, and EUA futures prices during the pilot trading period from the more liquid ECX. All the data for the Kyoto commitment period (spot and futures prices) are from PointCarbon. For the pilot trading period we consider spot and futures prices from October 4, 2005 to November 29, 2007, for the Kyoto commitment period for the period April 8, 2008–December 31, 2012. Spot contracts for EU emission allowances have a contract volume of one metric ton of CO$_2$ and are quoted in euros with a precision of two decimal places. For the pilot trading period, beginning on January 1, 2005, we consider 2006, 2007, and 2008 futures contracts; for the first Kyoto commitment period, beginning on January 1, 2008, we consider 2008, 2009, 2010, 2011, and 2012 futures; for the second Kyoto commitment period beginning on January 1, 2013 we consider 2013, 2014, and 2015 futures

contracts. The contract volume amounts to 1,000 tons of CO_2, and the contracts expire on the last business day in November (for the EEX futures) or on the last business day in December (for the ECX futures). For every futures contract, a settlement price in accordance with the current spot market price is established daily. According to a daily balancing of profit and loss (variation margin), the change in the value of a futures position is credited to the trading participant or is debited from the participant in cash. For both markets, delivery of the EU emission allowances is carried out up to two business days after maturity of the contracts. For the risk-free rates we use daily European Central Bank quotes for AAA-rated euro-area central government bonds. These quotes are available for bonds with a maturity ranging from 3 months to 5 years. We use linear interpolation to match the yields for different time horizons until maturity of the considered futures contracts. Whereas Madaleno and Pinho (2011), Gorenflo (2013), and Chang et al. (2013) assume a constant average risk-free rate, we use the observed daily rates for each maturity.

The pilot trading period

We begin with an analysis of the relationship between spot and futures prices for the pilot trading period. Whereas spot trading started already in January 2005, when the EU-wide CO_2 emissions trading system went into operation, futures contracts have been traded only since October 2005.

As was illustrated in chapter 1 (figure 1.5), at the commencement of trading EUA spot prices increased significantly from an initial price level of €8 up to almost €30 in the first six months. Among the reasons for this were constantly high gas and oil prices in the United Kingdom, relatively low world coal prices, and an extremely dry summer in the southwest of Europe. Because of the high temperatures and the drought, hydro-storage plants could not be fully utilized, especially in Spain. In addition, a shortage of cooling water for nuclear power plants led to higher utilization of coal-fired power plants and therefore increased the demand for CO_2 permits. Spot prices peaked at €29.21 on July 11, then fell back to about €22 in August, remaining there until the end of 2005. Extremely cold weather in January 2006 led to a substantial increase in allowance prices, which reached €29.78 on April 18, 2006.

Shortly after the April 2006 peak, news spread that a number of participating countries had given their industries such generous emission caps that there was no need for them to reduce their emissions.

On April 25, the Netherlands and the Czech Republic announced that their emissions were 7 percent and 15 percent below the respective allocations. Prices fell from €29.37 on April 24 to €9.13 on May 12. Spot prices then rose to approximately €18 by the end of May. After that, a more or less continuous decrease in spot prices until the end of the trading period could be observed. By the beginning of January 2007 the prices had already decreased to approximately €5, and by the end of March 2007 prices had fallen below €1 for the first time. They then declined steadily. On the last trading day (December 28, 2007), a price of €0.02 was observed.

To investigate the relationship between spot and futures allowance prices, we consider the period that began on October 4, 2005, when the trading of futures contracts commenced at the EEX. Figure 8.1 displays pilot period spot prices and futures prices for contracts with delivery in November 2007 and November 2008. Although there is a strong similarity between spot prices and 2007 futures prices, there is clearly less co-movement between the spot market and futures contracts with maturity during the first Kyoto period. Note that no futures contracts for the second Kyoto commitment period were traded during Phase I.

Table 8.2 reports the correlation coefficients between daily returns of emission allowances' spot and futures prices for the period from October 4, 2005 to April 24, 2006—that is, before the news of potential over-allocation spread. Table 8.3 reports correlations for the period

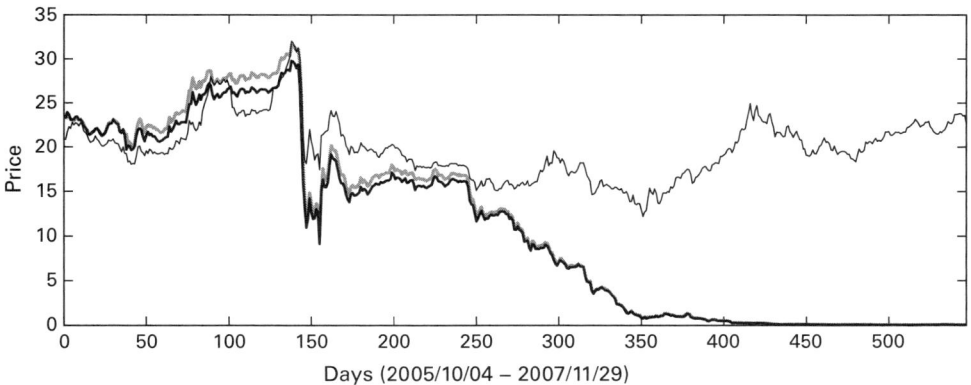

Figure 8.1
EUA spot price (solid) and futures prices for delivery in 2007 (dashed) and 2008 (dotted) for October 4, 2005 to November 29, 2007. Because the time scale uses business days, there are approximately 250 observations per year.

Table 8.2
Correlations between returns from spot and futures contracts for the pilot period (2006, 2007) and the Kyoto commitment period (2008–2012) for market quotes from October 4, 2005 to April 24, 2006.

Delivery	Spot	2006	2007	2008	2009	2010	2011	2012
Spot	1	0.8715	0.8596	0.6259	0.6147	0.5882	0.5800	0.5733
2006		1	0.9898	0.7126	0.7034	0.6732	0.6637	0.6558
2007			1	0.7170	0.7109	0.6812	0.6716	0.6631
2008				1	0.9834	0.9462	0.9361	0.9202
2009					1	0.9608	0.9506	0.9371
2010						1	0.9939	0.9840
2011							1	0.9963
2012								1

Table 8.3
Correlations between returns from spot and futures contracts for the pilot period (2006, 2007) and the Kyoto commitment period (2008–2012) for market quotes from April 25, 2006 to August 2, 2007.

Delivery	Spot	2006	2007	2008	2009	2010	2011	2012
Spot	1	0.9761	0.9617	0.5395	0.5467	0.5447	0.5369	0.5392
2006		1	0.9979	0.8424	0.8528	0.8518	0.8500	0.8438
2007			1	0.5723	0.5758	0.5726	0.5724	0.5734
2008				1	0.9922	0.9809	0.9750	0.9646
2009					1	0.9880	0.9817	0.9720
2010						1	0.9927	0.9851
2011							1	0.9918
2012								1

from April 25, 2006 to August 2, 2007. Let us first consider the time period before the significant drop of spot prices in April and May 2006. The results confirm the observation of figure 8.1: there is a strong correlation between the returns of spot prices and pilot-period futures prices, yielding $\rho > 0.8$ for futures with delivery in 2006 and 2007. The correlation between spot returns and returns of futures contracts for the Kyoto period is clearly lower but still significant, yielding correlations around 0.6. The lower correlations between Phase I spot prices and Phase II futures contracts can be attributed to the fact that no banking of allowances from the pilot period to the Kyoto commitment period was allowed such that pilot period allowances could not be used in the years 2008–2012. In general, the correlation is slightly decreasing with maturity, indicating that opinions of investors of a distant future environment are less affected by short-term price movements. Hence, we also find some evidence of the Samuelson or time-to-maturity effect. Further, we observe that the returns of futures contracts for the same trading period—either the pilot period or the Kyoto period—also show very high correlations. For the pilot period, we get a correlation between 2006 and 2007 futures returns of approximately $\rho = 0.99$; for the Kyoto commitment period, the correlations are between 0.92 and 0.99.

Correlations between contracts referring to Phase I and Phase II are lower when considering data that also includes observations during and after the significant drop in April and May of 2006. For this analysis we include spot prices until August 2, 2007, the first time the spot price fell below €0.10. The reason for this is that after the beginning of August 2007 there were hardly any changes in the spot price. Further, as a result of the €0.01 tick size of trading, returns in the spot market were at least of a magnitude of 10 percent, which may have biased the results of the correlation analysis. Further, note that the 2006 futures contract expired on November 29, 2006, such that for this contract estimated correlations are based entirely on the period from April 25, 2006 to November 29, 2006. Therefore, correlations between this contract and Kyoto-commitment-period contracts are still fairly high. Correlations between returns of spot and futures returns within the same trading period remain high, while the correlation between spot and Kyoto period futures returns have dropped. They are approximately 0.53 for returns from Phase I spot contracts and Kyoto commitment period futures contracts, and around 0.57 for returns from the 2007 futures contract and Kyoto commitment futures contracts. These results indicate that after the news of relatively high allocation of allowances the price signal given by prices from contracts of the Kyoto commitment period was less relevant for the pilot trading period and vice versa. Market participants were aware that new allocation plans would be created for Phase II with allocations below those of the pilot trading period. Further, the fact that banking of allowances from the pilot period to the Kyoto commitment period was prohibited made Phase I and Phase II allowances essentially two different assets.

Figure 8.2 displays the term structure of emission allowance spot and futures prices with yearly maturities from November 2006 to November 2012. For each trading day in October 2005, January 2006,

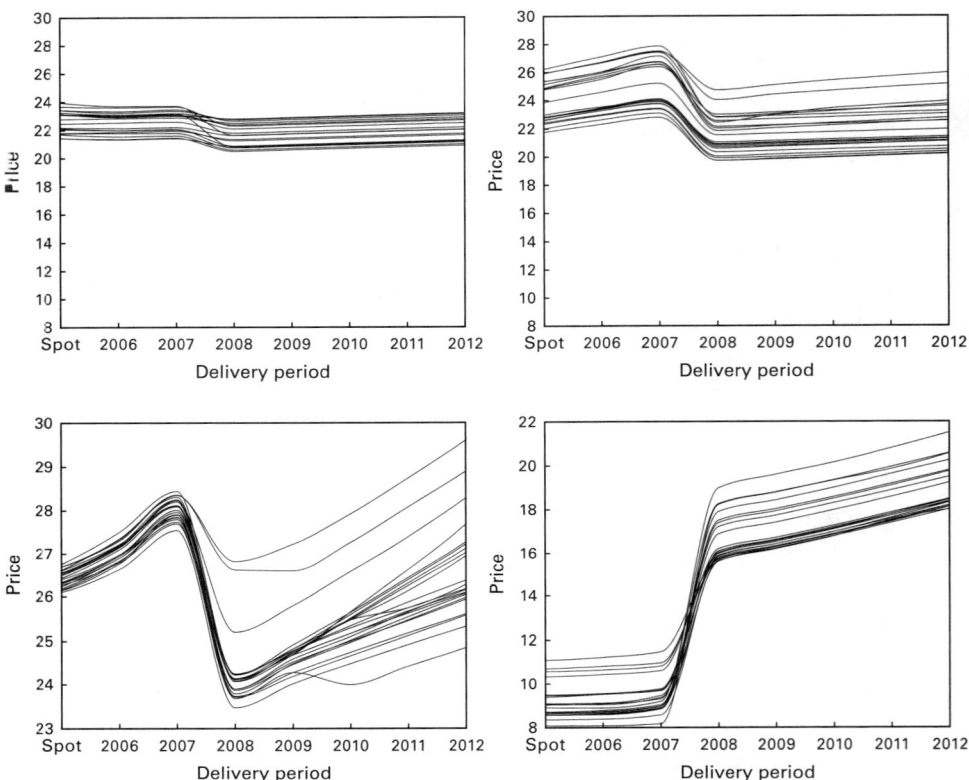

Figure 8.2
Term structure for spot and futures prices for each trading day of October 4–31, 2005
(upper left), January 1–31, 2006 (upper right), March 1–31, 2006 (lower left), and November 1–30, 2006 (lower right).

March 2006, and November 2006 the daily observed spot and futures
prices are connected by a smoothed line using cubic interpolation. We
find that the term structure of futures prices is dynamic and shows
quite different behavior through time. During the initial trading period,
in October 2005, futures prices for both the pilot period and the Kyoto
period were slightly below current spot prices. While there was a quite
flat term structure for the pilot period, a slightly increasing term structure of futures prices could be observed for the Kyoto commitment
period. In January 2006, for the pilot period an increasing term structure can be observed, whereas the term structure for the Kyoto period
is only slightly increasing. Futures prices for the Kyoto commitment
period are still below the spot price and futures prices of the pilot

period. In May 2006, after the news of over-allocation of emission rights in a number of European countries was published, futures prices for the Kyoto period are slightly higher than the spot and pilot period futures prices. In September 2006, a clearly increasing term structure can be observed and futures prices for the Kyoto period are significantly above the spot and pilot period futures prices. We conclude that starting from May 2006 the relationship between pilot-period spot prices and Kyoto-commitment-period futures prices showed significant changes. Though the spot prices and the Phase I futures prices fell significantly after the news of over-allocation of allowances, Kyoto period futures contracts were clearly less affected by that news. The latter can be attributed to the fact that no banking of allowances from the pilot period to the Kyoto commitment period was allowed. Further, market participants were aware that new allocation plans would be provided for the Kyoto commitment period that probably would be below allocations for the pilot trading period.

Figure 8.3 displays the volatility term structure for spot and futures prices with delivery in November 2006 until November 2012. According to the Samuelson effect, we would expect a declining term structure of the forward price volatility. Obviously, the volatility term structure of spot and futures prices also shows strong dynamics through time. Over the whole period from October 4, 2005 until September 29, 2006, the volatility of futures for the pilot period was higher than the spot price volatility, whereas for the Kyoto commitment period lower volatilities in futures prices could be observed. Quite different results are obtained if subperiods are examined. For the first three months of trading (from October to December 2005), a decreasing volatility term structure for the pilot period can be observed. For the Kyoto period the volatility term structure was flat, however, futures prices showed significantly higher volatilities. From January to March 2006 there is a monotonic increasing volatility term structure in futures prices. The lowest volatility can be observed for the spot price; the highest volatility is exhibited by the 2012 futures. A quite opposite behavior can be found for the period from April through June. After the news of over-allocation in certain countries was published, spot prices and pilot-period futures prices showed strong reaction and exhibited extreme volatilities in comparison to the first six months of futures trading. The standard deviation on daily prices rose from approximately $\sigma = 6$ for the spot prices to approximately $\sigma = 6.5$ for the 2007 futures. Further, for the pilot period the volatility term structure is increasing. Kyoto

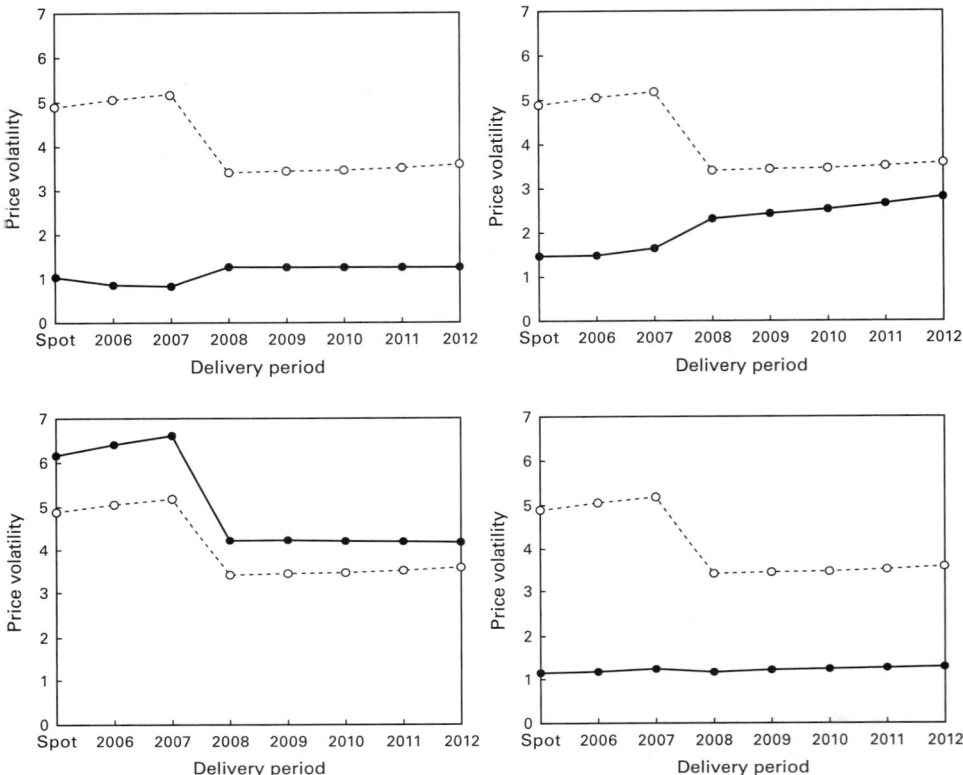

Figure 8.3
Volatility for spot and futures prices with delivery in 2006–2012. In all panels, the dotted line represents the volatilities for the whole period (October 4, 2005–September 29, 2006). Bold lines represent the volatilities for the trading period October 4, 2005–December 31, 2005 (upper left), January 2, 2006–March 31, 2006 (upper right), April 3, 2006–June 30, 2006 (lower left) and July 1, 2006–September 29, 2006 (lower right).

period futures prices showed less reaction to the news and clearly less volatility. Here the term structure remains flat, the standard deviation of daily prices is approximately $\sigma = 4.2$ for all futures. For the last three months the volatility term structure is slightly increasing but quite flat. For all traded products the standard deviation of daily prices is very close to $\sigma = 1$. Overall, the results contradict other studies in the literature on the volatility of futures prices and give ambivalent results on the Samuelson effect. Although we found that the correlation between spot and futures prices decreases with longer maturity of the futures, separately examining the volatility of futures prices for the pilot period and the Kyoto period we find a rather increasing term structure and strong dynamics through time.

We will now investigate the behavior of the convenience yield of CO_2 emission allowance futures prices for the pilot trading period. For the risk-free rates we use daily European Central Bank quotes for AAA-rated eurozone government bonds, using linear interpolation to match the yields for different time horizons until maturity of the futures contracts. Recall that under the standard cost-of-carry approach we would expect the convenience yield to be zero, and thus that the equation $F_{(t,T)} = e^{r(T-t)}S_t$ would hold. Given equation 3, in the following we consider absolute values of the convenience yield as

$$\gamma_{(T-t)} = S_t e^{r(T-t)} - F_{t,T}.$$

Figure 8.4 displays the convenience yield for the pilot period futures contract with delivery in November 2006 and November 2007. We find that the convenience yield was initially significantly different from zero for both contracts. This confirms results by other studies—for example, Milunovich and Joyeux (2010) report that none of the pilot period carbon futures contracts are priced according to a cost-of-carry model relationship. Milunovich and Joyeux argue that the mis-pricing could

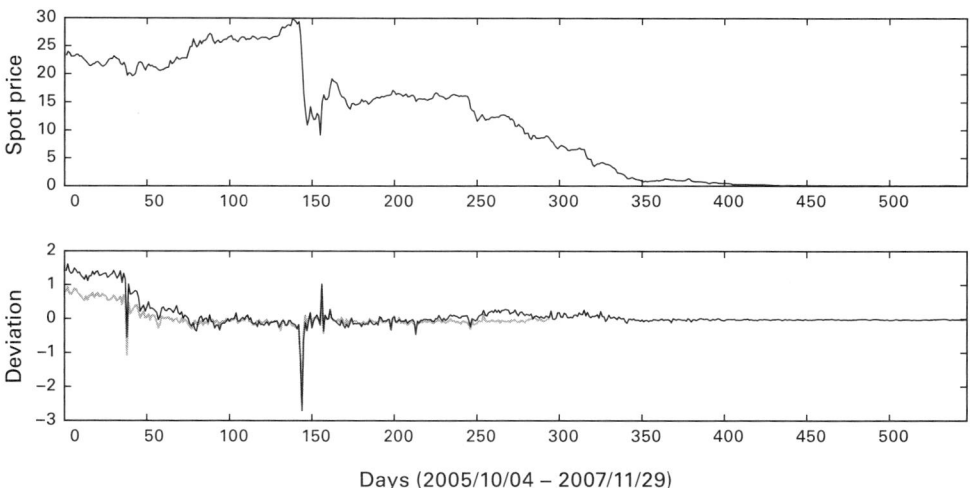

Days (2005/10/04 – 2007/11/29)

Figure 8.4
Upper panel: Spot prices (euros per ton) from October 4, 2005, to November 29, 2007. Lower panel: Convenience yields (euros per ton) for 2006 (dashed gray) and 2007 (solid) EUA futures.

be due to a large standard error associated with the estimated parameter on the interest rate variable. We can also observe three substantial shocks or rather short-lived spikes in the convenience yield time series, indicating a reaction of the spot/futures price relationship to market news. The most significant one is observed in April 2006, when, because the news of relatively high allocation of permits, the convenience yield suddenly became negative as a consequence of the spot price's substantial fall, whereas 2006 and 2007 futures prices remained higher for a short period of time. The closer we get to the end of the pilot trading period, the smaller the convenience yield becomes. Closer to the end of the pilot trading period, the price for the 2007 futures contract also approaches zero.

Overall, our findings suggest that, insofar as banking and borrowing of allowances were allowed during the pilot period, initially there were potential arbitrage opportunities in the market for carbon permits. Since at least for the first 6 months of trading the convenience yields were significantly different from zero and none of the futures contracts followed a cost-of-carry relationship with the spot price, market participants should have been able to achieve riskless profits by trading.

Quite different results can be obtained for the relationship between Phase I spot contracts and Kyoto-commitment-period futures contracts. Note, however, that, because of new allocation plans for the Kyoto commitment period, the term "convenience yield" is not really appropriate, since pilot-period and Kyoto-period contracts refer to different trading periods and thus also to products that are subject to different levels of scarcity. Since banking of pilot-trading-period allowances for use during the Kyoto commitment period was prohibited, there was no immediate benefit of holding pilot period spot contracts with regard to the Kyoto commitment period. In what follows, we will refer to the difference between the spot and discounted futures price as deviation from the cost-of-carry relationship; however, we are aware that the use of this term might be misleading. As indicated by figure 8.5, which shows the observed deviation for 2008 and 2012 futures contracts, during the first six months of trading the yield was clearly positive, with values $0 < \gamma < 10$. After the news of sufficient allocation of allowances for the pilot period became known, initially the price shock on allowances obviously affected Kyoto-period futures contracts similar to the ones for the pilot trading period. However, the persistence of the shock on different futures contracts was of an entirely different nature: for the 2006 and 2007 futures, after a very short period with negative

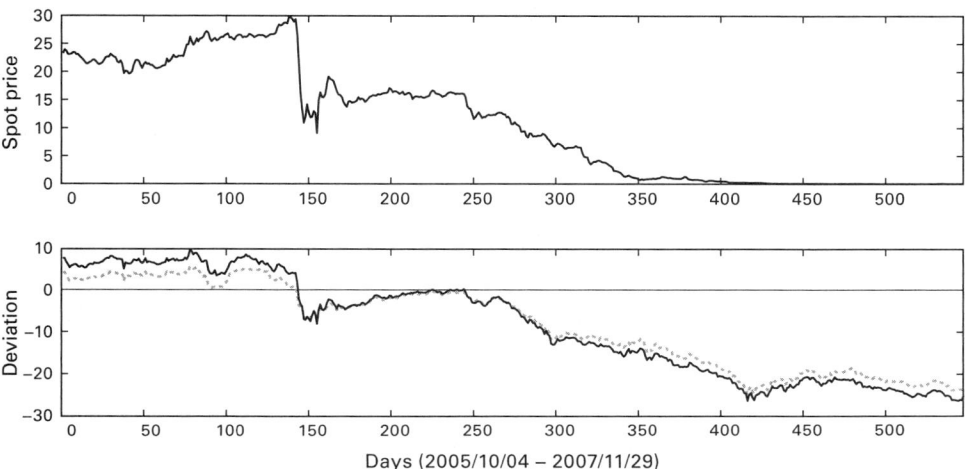

Figure 8.5
Upper panel: Spot prices (euros per ton) from October 4, 2005, to November 29, 2007.
Lower panel: Deviation from cost-of-carry relationship (euros per ton) for 2008 (dashed gray) and 2012 (solid) EUA futures contracts.

yields of approximately −2.5, the futures prices for the pilot period adapted to the price change quickly, and convenience yields approached zero. On the other hand, for Kyoto-period futures contracts the effect of the price shock on futures prices was not as dramatic as for the pilot period. The prohibition of banking between Phase I and Phase II and the expectation that there would be new national allocation plans for the Kyoto commitment period kept futures prices higher (between €12 and €25) until the end of the pilot trading period in 2007. Thus, as illustrated in figure 8.5, the deviation from the cost-of-carry relationship for the 2008–2012 futures contracts became significantly negative. As the price of the spot contract approaches zero, it basically equals −1 times the futures price; compare figures 8.1 and 8.5. Overall, the analysis of pilot-period spot prices and Kyoto-commitment-period futures prices reveals the following relationship: Although in the beginning pilot period spot prices were also considered as an indication for Kyoto-commitment-period allowance prices, after the news of relatively high allocation of permits the importance of Phase I prices for Phase II futures prices fell dramatically. Significantly higher prices for the Kyoto-commitment-period futures contracts indicate that market participants saw no privilege in holding the spot contract with respect to future periods. The major reasons for this were the prohibition of

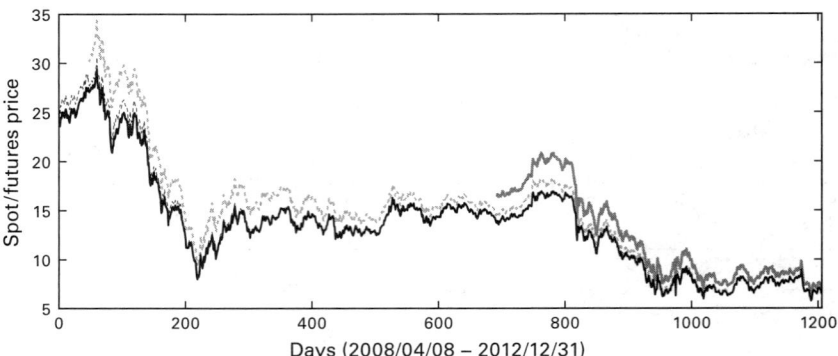

Figure 8.6
Spot price (solid black), December 2010 (dashed), December 2012 (dotted) and December 2014 (solid gray) futures price for the first Kyoto commitment period (April 8, 2008–December 31, 2012). The December 2010 futures contract expired on December 20, 2010 and the December 2012 futures contract on December 17, 2012. The first price observation for the 2014 futures contract was available on December 21, 2010.

banking between the pilot period and the Kyoto commitment period and market participants' expectations of lower allocations of allowances for the first Kyoto commitment period.

The Kyoto commitment period
We will now consider the relationship between spot and futures contracts for the Kyoto commitment period using data from April 8, 2008 to December 31, 2012. Figure 8.6 shows the spot price series as well as the December 2010, 2012, and 2014 futures prices for the period considered.[1] We observe that the Phase II EUA spot price (bold solid line) on April 8, 2008 was €23.53 and initially increased to its maximum level of €29.38 on July 1, 2008. What followed was a relatively rapid decline in prices down to €8.00 on February 12, 2009, which can mainly be attributed to the effects of the financial crisis and lower expectations about economic output in the eurozone due to the crisis. Spot prices increased again up to a level of €15.45 in May 2009 and remained between €13 and €16 till June 2011. Since then, owing to the European Sovereign Debt crisis, expectations about low economic output in future periods and the expectations of a relatively high allocation of allowances, prices fell to about €6.50 in December 2012. We also observe that spot and futures prices move simultaneously during the entire period.

Table 8.4
Correlations between returns from spot and futures contracts (2008–2015) for Kyoto-commitment-period market quotes from April 8, 2008 to December 31, 2012. Note that correlation coefficients between returns from the 2008 and 2013, 2014 and 2015 futures contracts could not be calculated because the 2008 contract expired before quotes for these contracts were available. The same is true for the correlation coefficient between 2009 and 2014, 2015 contracts and for 2010 and 2015 futures contracts.

Delivery	Spot	2008	2009	2010	2011	2012	2013	2014	2015
Spot	1.0000	0.9815	0.9915	0.9809	0.9708	0.9689	0.9897	0.9888	0.9580
2008		1.0000	0.9844	0.9732	0.9649	0.9618	-	-	-
2009			1.0000	0.9895	0.9797	0.9794	-	-	-
2010				1.0000	0.9870	0.9779	0.9784	-	-
2011					1.0000	0.9774	0.9863	0.9819	-
2012						1.0000	0.9955	0.9944	0.9626
2013							1.0000	0.9970	0.9667
2014								1.0000	0.9680
2015									1.0000

The co-movement of spot and futures contracts during the first Kyoto commitment period is also confirmed by looking at the correlation coefficients between returns from spot contracts and December 2008, 2009, ... , 2015 futures contracts in table 8.4. We find that correlations between spot returns and futures returns are all well above 0.95 and close to 1. This is also true for contracts referring to different trading period, i.e., Phase II and Phase III of the scheme. Therefore, results are quite different than what we had observed for contracts referring to the pilot period and the first Kyoto commitment period, in which inter-period banking was not allowed. Apparently, the fact that banking from Phase II to Phase III is allowed created a more similar behavior of returns for contracts referring to either period.

In figure 8.7 we plot the observed convenience yields for Kyoto commitment spot and December 2009, 2011, and 2012 futures contracts.[2] Similar to the pilot trading period, the market started in backwardation, with positive convenience yields indicating that the spot price was above the discounted price of Kyoto-commitment-period futures contracts. In the course of time, the market situation changed from backwardation to contango for the first time in July 2008. Prices were approximately in line with the cost-of-carry relationship until end of October 2008, but after that month the convenience yield for the contract was negative. During the period from March to December 2009,

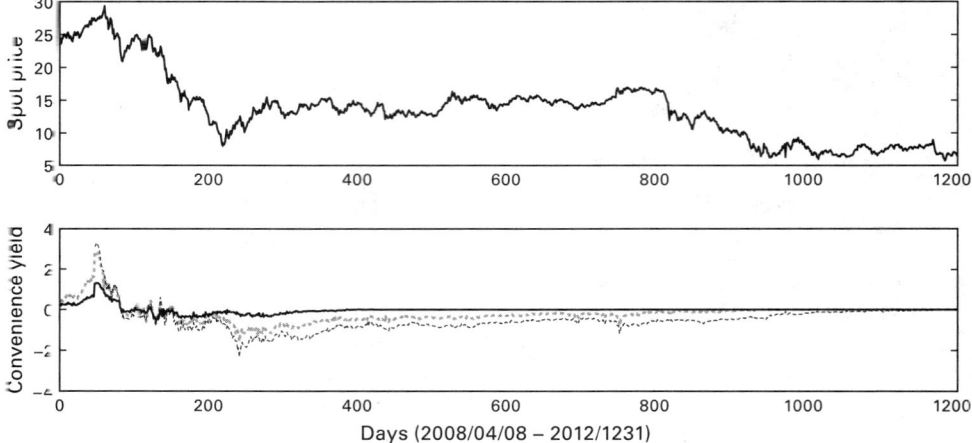

Figure 8.7
Upper panel: Spot prices (euros per ton) from April 8, 2008 to December 31, 2012. Lower panel: Convenience yields (euros per ton) for 2009 (solid), 2011 (dashed) and 2012 (dotted) EUA futures contracts. The 2009 futures contract expired on December 14, 2009, the 2011 contract on December 19, 2011, and the 2012 contract on December 17, 2012.

the convenience yield for 2011 and 2012 futures contracts was significantly smaller than zero. Overall, we find that similar to the pilot period none of the spot or futures contracts were priced according to the cost-of-carry relationship. The effect is more pronounced for futures contracts with longer maturities, such as contracts maturing in December 2011 or in December 2012. As the contracts get closer to the expiry date, the convenience yield becomes smaller and approaches zero. The negative convenience yields for Kyoto period futures contracts from 2009 on indicate that market participants saw no advantage in holding the allowance now with respect to future periods. Because banking and borrowing were allowed during the Kyoto commitment period (2008–2012), one might attribute the deviation from the cost-of-carry relationship to a large standard error associated with the estimated parameter on the interest rate variable or to different market expectations about interest rates in coming years. Recall that in the financial crisis risk-free rates in the eurozone fell from around 4 percent in September 2008 to 0.4 percent in September 2009, and that they have remained very low since then.

Figure 8.8 displays the results for the relationship between Phase II spot and Phase III futures contracts. Because inter-phase banking is

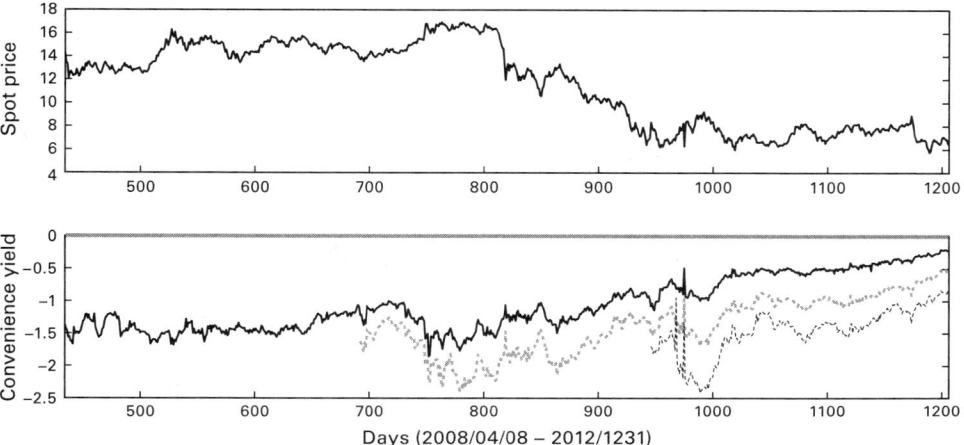

Figure 8.8
Upper panel: Spot prices (euros per ton) from December 16, 2009 to December 31, 2012. Lower panel: Convenience yields (euros per ton) for December 2013 (solid), December 2014 (dashed) and December 2015 (dotted) EUA futures contracts. Phase III futures contract prices were available from December 16, 2009 (2013 futures), December 21, 2010 (2014 futures) and December 20, 2011 (2015 futures).

allowed, the EU ETS enables market participants to use Phase II permits also during Phase III. On the other hand, borrowing of permits from Phase III and using the allowances in Phase II is not allowed. Note that prices for the considered Phase III futures contracts were available only from December 16, 2009 (2013 futures), December 21, 2010 (2014 futures), and December 20, 2011 (2015 futures). Therefore, figure 8.8 provides a plot of the spot price and observed convenience yields only for that time period. We find highly negative convenience yields for all Phase III futures contracts, usually in the range between −1 and −3 for December 2009 until the end of 2011. During 2012, observed convenience yields were smaller; they remained clearly below zero for all contracts.

Note that our findings with respect to a clear deviation from the cost-of-carry relationship for Phase II are in line with earlier studies examining the relationship between EUA spot contracts and futures contracts during the Kyoto commitment period (Madaleno and Pinho 2011; Gorenflo 2013; Chang et al. 2013). However, the consistently negative sign of observed convenience yields from March 2009 on contradicts results reported in these studies, at least partially. Though none of the earlier studies examines the relationship for the entire Phase II,

Madaleno and Pinho (2011) and Chang et al. (2013) report positive convenience yields during late 2009 and 2010 for some of the futures contracts (usually contracts with maturity during Phase II, i.e., expiry in December of 2010, 2011, or 2012). The main reason for the deviation in our results may be different assumptions about the risk-free rate. Whereas Madaleno and Pinho (2011) assume a constant interest rate for the estimation period of 4 percent, Chang et al. (2013) choose a constant free-risk rate equal to the average coupon rate of 3.06 percent, i.e., the rate for three-year government bonds issued in 2010 in the European Union. Gorenflo (2013) states that in his analysis the interest rate is assumed to be constant over time. Note that in our study we relax the assumption of a constant average risk-free rate and use actual daily European Central Bank quotes for AAA-rated eurozone central government bonds for different maturities. Further, to match the yields for different time horizons until maturity of the considered futures contracts we use linear interpolation between quoted interest rates. As mentioned earlier, risk-free rates in the eurozone have fallen from around 4 percent in September 2008 to below 1 percent since late September 2009. Therefore, it is no surprise that in our analysis we obtain different results in comparison to previous studies, where a significantly higher interest rate has been applied.

There are several things that could explain the negative convenience yields and the clear deviation from the cost-of-carry relationship for Phase II. The first may be the extremely low risk-free rates in the eurozone from 2009 on. That decrease from 4 percent to rates near 0.5 percent was initially due to the financial crisis, but the yields of AAA-rated government bonds have remained low ever since. Clearly, as is indicated by equations 3 and 4, the risk-free rate plays a substantial role in the cost-of-carry model and, therefore, also for quantifying deviations from this relationship and the calculation of the convenience yield. We also observe that convenience yields become more significant once risk-free rates in the eurozone drop to the low levels we have seen since 2009. During periods of very low interest rates, negative convenience yields for risky assets may be more likely, perhaps as a result of market expectations about rising interest rates in coming periods.

Second, the significantly negative convenience yields for Phase II and Phase III futures contracts indicate that long positions in futures contracts are priced higher than would be suggested by the simple cost-of-carry relationship. Generally, a contango market as it is observed during Phase II would suggest currently available supply but potential medium-to-long-term shortages of a commodity. Under such a scenario, consumers might be interested in buying insurance against rising prices in the futures market. Therefore, a greater interest in long futures positions will drive prices of these contracts up to a level that is higher than what may be suggested by the cost-of-carry relationship. Therefore, observed negative convenience yields may be interpreted as consumers' willingness to pay an additional premium for a hedge against rising prices or a future shortage of EUAs. Clearly, it can also be interpreted as a hedge against potential changes in regulation that may reduce the availability of permits in coming years.

The last explanation refers to the possibility of banking EUAs and the surplus of allowances available during Phase II. Generally, the theory of storage would suggest a negative relationship between the convenience yield and inventory—see, e.g., Pindyck 2001. The owner of a commodity, who is free to consume it until maturity, is prepared for unexpected shortages in supply or increases in demand. The convenience yield then represents this additional benefit of holding a unit of inventory, for instance, to be able to meet unexpected demand. The value of this benefit should then be negatively related to the level of inventory. One could argue that it is particularly high if inventories of a commodity are low and consumers are forced to secure a short-term supply. On the other hand, high levels of inventory will reduce the benefits and, therefore, also the convenience yield. Considering the continuously increasing level of surplus allowances during the first Kyoto commitment period[3] and the extensive use of external credits coming from two Kyoto Protocol mechanisms—the Clean Development Mechanism and Joint Implementation, one could argue that throughout Phase II an increasingly higher level of inventory was accumulated. Therefore, the change in the market from backwardation to contango and significantly negative convenience yields for futures contracts could be a result of an increasing level of surplus allowances and banking.

Figure 8.9 illustrates the volatility term structure for spot and futures prices with delivery in December 2009 until December 2014. According to the Samuelson effect we would expect a declining levels of volatility for futures contracts with longer maturities. Obviously, also the volatility term structure of spot and futures prices shows strong dynamics through time. During the period from July 1, 2008 to September 30, 2008, the volatility of futures contracts for Phase II and Phase III was

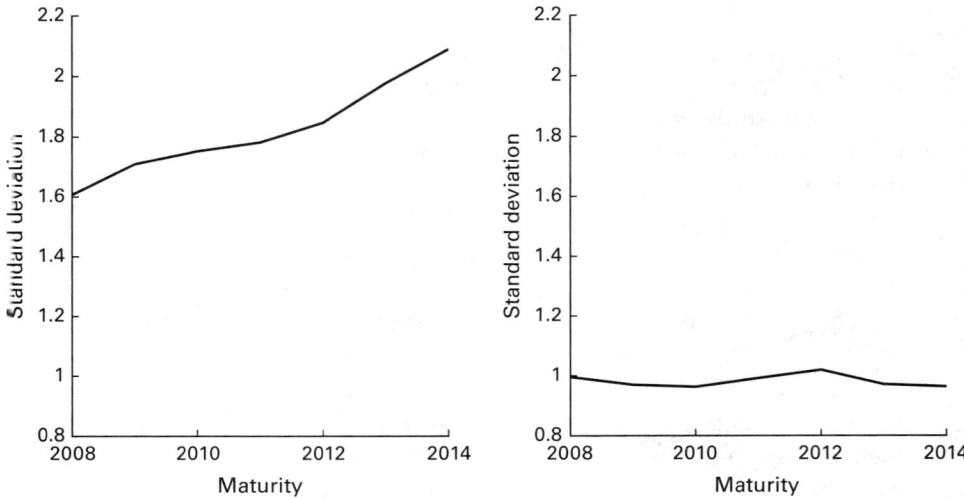

Figure 8.9
Volatility term structure of daily prices for the considered spot and 2009–2014 December futures contracts for the trading period July 1, 2008–September 30, 2008 (left) and April 1–June 30, 2009 (right).

greater than that of spot prices. The results were quite different in the period from April 1 to June 30, 2009, when the volatility term structure was quite flat. In some other subperiods, a decreasing volatility term structure could be observed. Overall, we find a rapidly changing behavior of the volatility term structure that often contradicts the Samuelson effect. In fact, for many periods the volatility of futures contracts with longer maturities is significantly greater than the volatility of spot prices. Overall, EUAs seem to exhibit behavior suggesting that further investigation of spot and futures CO_2 emission allowance prices is warranted.

Conclusion

We find that the price behavior of emission allowances in the spot and futures market is quite different from those of other commodity markets. We observe a quite dynamic behavior of the term structure both for allowance prices and volatilities. Whereas in general correlations between spot and futures prices decrease with time to maturity, the term structure of EUA prices shows significant changes through time. We find that the market has changed from initial backwardation to contango both for the pilot trading and for the Kyoto commitment period. Thus, futures prices are higher than the current spot price and deviate from the standard cost-of-carry relationship. Also, the term structure of volatilities for spot and futures prices is subject to several changes. We find an overall increasing price volatility with maturity of the contracts for both periods. This somehow contradicts the time-to-maturity or Samuelson effect, which suggests a typically declining term structure in the volatility of futures prices as maturity increases.

Furthermore, the observed convenience yields in futures contracts are significantly different from zero, particularly for contracts with longer maturities. Considering the first Kyoto commitment period (2008–2012), we find that the market has changed from initial backwardation to contango, with significantly negative convenience yields in futures contracts. We suggest three main reasons for this relatively large and persistent deviation from the cost-of-carry relationship. The first one may be the extremely low risk-free rates in the eurozone from 2009 on. The decrease in these rates from 4 percent to about 0.5 percent was initially due to the financial crisis, but the yields of AAA-rated government bonds have remained low ever since. The second explanation refers to the fact that market participants are interested in buying insurance against rising prices and therefore may be willing to pay an additional premium in the futures market for a hedge against changes in regulation or a future shortage of EUAs that would increase permit prices. The last explanation refers to the increasing level of surplus allowances and banking during the first Kyoto commitment period. In view of the negative relationship between convenience yields and the level of inventory, this may also explain the significant negative convenience yields during Phase II. We recommend a more thorough investigation of the relationship between observed convenience yields in the CO$_2$ allowance futures market and the suggested factors.

Acknowledgments

This research was supported by the Australian Research Council through grant no. DP1096326, by the Deutsche Forschungsgemeinschaft through grant no. SFB 649, by the Croatian Science Foundation under grant no. IP-2013-11-2203, and by the Robert Schumann Centre at European University Institute (EUI).

210

Notes

1. Note that the 2010 futures contract expired on December 20, 2010, the 2012 futures contract on December 17, 2012, whereas the first price observation for the 2014 futures contract was available on December 21, 2010.

2. Note that the 2009 futures contract expired on December 14, 2009, the 2011 contract on December 19, 2011 and the 2012 contract on December 17, 2012.

3. See, e.g., http://europeanclimatepolicy.eu/.

References

Benz, E., and J. Hengelbrock. 2008. Liquidity and price discovery in the European CO_2 futures market: An intraday analysis. Working paper, Bonn Graduate School of Economics.

Benz, E., and S. Trück. 2006. CO_2 emission allowances trading in Europe—specifying a new class of assets. *Problems and Perspectives in Management* 3: 4–15.

Benz, E., and S. Trück. 2008. Modeling the price dynamics of CO_2 emission allowances. *Energy Economics* 31 (1): 4–15.

Bierbrauer, M., C. Menn, S. T. Rachev, and S. Trück. 2007. Spot and derivative pricing in the EEX power market. *Journal of Banking & Finance* 31 (11): 3462–3485.

Bodie, Z., and V. Rosansky. 1980. Risk and return in commodities futures. *Financial Analysts Journal* 36: 27–39.

Bokenkamp, K., H. LaFlash, V. Singh, and D. Wang. 2005. Hedging carbon risk: Protecting customers and shareholders from the financial risk associated with carbon dioxide emissions. *Electricity Journal* 18 (6): 11–24.

Botterud, A., T. Kristiansen, and M. Ilic. 2010. The relationship between spot and futures prices in the Nord Pool electricity market. *Energy Economics* 32 (5): 967–978.

Bredin, D., and C. Muckley. 2011. An emerging equilibrium in the EU emissions trading scheme. *Energy Economics* 33: 353–362.

Carmona, R., and J. Hinz. 2011. Risk-neutral models for emission allowance prices and option valuation. *Management Science* 57 (8): 1453–1468.

Chang, E. 1985. Returns to speculators and the theory of normal backwardation. *Journal of Finance* 40: 193–208.

Chang, K., S. Sheng Wang, and K. Peng. 2013. Mean reversion of stochastic convenience yields for CO_2 emissions allowances: Empirical evidence from the EU ETS. *Spanish Review of Financial Economics* 11 (1): 39–45.

Chesney, M., and L. Taschini. 2012. The endogenous price dynamics of emission allowances and an application to CO_2 option pricing. *Applied Mathematical Finance* 19 (5): 447–475.

Chevallier, J. 2009a. Carbon futures and macroeconomic risk factors: A view from the EU ETS. *Energy Economics* 31 (4): 614–625.

Chevallier, J. 2009b. Modelling the convenience yield in carbon prices using daily and realized measures. *International Review of Applied Financial Issues and Economics* 1 (1): 56–73.

Chevallier, J. 2011. Detecting instability in the volatility of carbon prices. *Energy Economics* 33 (1): 99–110.

Conrad, C., D. Rittler, and W. Rotfuss. 2012. Modeling and explaining the dynamics of European Union allowance prices at high frequency. *Energy Economics* 34 (1): 316–326.

Daskalakis, G., D. Psychoyios, and R. Markellos. 2009. Modeling CO$_2$ emission allowance prices and derivatives: Evidence from the EEX. *Journal of Banking & Finance* 33 (7): 1230–1241.

Diko, P., S. Lawford, and V. Limpens. 2006. Risk premia in electricity forward prices. *Studies in Nonlinear Dynamics & Econometrics* 10 (3), article 7.

Fama, E., and K. French. 1987. Commodity futures prices: Some evidence on forecast power, premiums, and the theory of storage. *Journal of Business* 60 (1): 55–73.

Fichtner, W. 2004. Produktionswirtschaftliche Planungsaufgaben bei CO$_2$-Emissionsrechten als neuen Produktionsfaktor. Habilitation, Universität Karlsruhe.

Geman, H. 2005. *Commodities and Commodity Derivatives.* Wiley.

Gibson, R., and E. Schwartz. 1990. Stochastic convenience yield and the pricing of oil contingent claims. *Journal of Finance* 45 (3): 959–976.

Gorenflo, M. 2013. Futures price dynamics of CO$_2$ emission allowances. *Empirical Economics* 45: 1025–1047.

Gronwald, M., J. Ketterer, and S. Trück. 2011. The relationship between carbon, commodity and financial markets: A copula analysis. *Economic Record* 87 (s1): 105–124.

Hicks, J. 1946. *Value and Capital,* second edition. Oxford University Press.

Hintermann, B. 2010. Allowance price drivers in the first phase of the EU ETS. *Journal of Environmental Economics and Management* 59 (1): 43–56.

Hull, J. C. 2005. *Options, Futures, and Other Derivatives,* sixth edition. Prentice-Hall.

Isenegger, P., R. Wyss, and S. Marquardt. 2013. The valuation of derivatives on carbon emission certificates—A GARCH approach. Working paper.

Keynes, J. 1930. *A Treatise on Money,* volume 2. Macmillan.

Longstaff, F., and A. Wang. 2004. Electricity forward prices: A high-frequency empirical analysis. *Journal of Finance* 59: 1877–1900.

Madaleno, M., and C. Pinho. 2011. Risk premia in CO$_2$ allowances: Spot and futures prices in the EEX market. *Management of Environmental Quality* 22 (5): 550–565.

Milunovich, G., and R. Joyeux. 2010. Market efficiency and price discovery in the EU carbon futures market. *Applied Financial Economics* 20 (10): 803–809.

Niblock, S., and J. Harrison. 2012. Do dynamic linkages exist among European carbon markets? *International Business and Economics Research Journal* 11 (1): 33–44.

Paolella, M., and L. Taschini. 2008. An econometric analysis of emission trading allowances. *Journal of Banking & Finance* 32 (10): 2022–2032.

Pindyck, R. 2001. The dynamics of commodity spot and futures markets: A primer. *Energy Journal (Cambridge, Mass.)* 22: 1–29.

Samuelson, P. 1965. Proof that properly anticipated prices fluctuate randomly. *Industrial Management Review* 6: 41–49.

Seifert, J., M. Uhrig-Homburg, and M. Wagner. 2008. Dynamic behavior of CO_2 spot prices—a stochastic equilibrium model. *Journal of Environmental Economics and Management* 56 (2): 180–194.

Uhrig-Homburg, M., and M. Wagner. 2009. Futures price dynamics of CO_2 emission allowances: An empirical analysis of the trial period. *Journal of Derivatives* 17 (2): 73–88.

Wei, S., and Z. Zhu. 2006. Commodity convenience yield and risk premium determination: The case of the US natural gas market. *Energy Economics* 28: 523–534.

Weron, R. 2006. *Modeling and Forecasting Electricity Loads and Prices: A Statistical Approach.* Wiley.

Weron, R. 2008. Market price of risk implied by Asian-style electricity options and futures. *Energy Economics* 30 (3): 1098–1115.

Weron, R., and M. Zator. 2014. Revisiting the relationship between spot and futures prices in the Nord Pool electricity market. *Energy Economics* 44: 178–190.

9   Trading Behavior in the EU ETS

Ralf Martin, Mirabelle Muûls, and Ulrich J. Wagner

In 2005, with the reduction of greenhouse-gas emissions becoming an increasingly important policy objective worldwide, the member states of the European Union took the lead by implementing the EU Emissions Trading System. This flagship policy instrument of the EU's climate action imposes an overall cap on emissions of carbon dioxide from power generators and large industrial emitters. The system provides a framework for emitters and other parties to trade permits.

According to economic theory, the cap-and-trade system ensures that the total costs of abating CO_2 emissions are minimized. The cap creates certainty about the total amount of emissions, but leaves uncertain how much complying with it will cost individual firms. Another important theoretical property of the cap-and-trade system is that the allocation of permits to firms, although it determines the distribution of rents that emerge from imposing scarcity on carbon emissions, should have no consequences for firms' abatement choices. The only parameter that a firm should take into consideration when deciding how much to curb its emissions and how much to invest in abatement technology is the expected price of an emission permit. The notion that abatement decisions are independent of the distribution of permits is referred to as the "independence property" of emissions trading (Montgomery 1972; Hahn and Stavins 2011). This property depends on firms understanding the functioning of a cap-and-trade system and acting rationally, on the basis of a comparison of the allowance price and their marginal abatement cost, in their abatement and trading decisions.[1]

However, even after several years of emissions trading, not much is known about how firms behave in the market for emission allowances. This lack of knowledge is particularly striking in the case of industrial emitters, which have been scrutinized less than the large power generators that are major players in the allowance market. The EU's

Community Independent Transaction Log (CITL) provides data on allowance trading, allocated allowances, and verified emissions for each year an installation participates in the EU ETS.[2] Ellerman and Buchner (2007) examine the patterns of allocations and verified emissions and the resulting short or long positions in permit holdings and find evidence suggesting that the power sector dominates the allowance market. Moreover, they note that most trades happen within countries.

The bulk of the literature, however, is based on surveys. Surveys have been a common tool in the empirical analysis of the EU ETS since its beginnings, which can be explained by the initial lack of official data suitable for qualitative and quantitative analyses. Empirical studies show that firms were trading mostly to buy the permits they needed to be in compliance. This finding, which has proved to be fairly robust, emerged as early as in the summer of 2005, when the EU Commission commissioned a mail survey of stakeholders. (See McKinsey and Ecofys 2006.) Among 144 industrial companies and power generators, 62 percent reported that they were not trading in the market and only 48 percent said they were pricing in the value of their allowances in their daily operations. However, 72 percent reported they were planning to include them in their future marginal production decisions.

Similarly, surveys of firms subject to the EU ETS (hereafter, EU ETS firms) in Germany (Löschel et al. 2010), in Sweden (Sandoff and Schaad 2009) and in the Carbon Disclosure Project (Pinkse and Kolk 2007) find evidence that firms act as if the EU ETS were a compliance mechanism rather than a market-based policy instrument. This type of behavior could have increased the overall cost of compliance, especially if firms with excess amounts of permits did not make them available to other market participants. In a recent paper, Jaraitė-Kažukauske and Kažukauskas (2014) analyze allowance transactions of all EU ETS firms during the first trading phase (2005–2007) and show that transaction costs are important in explaining lack of participation in the market in those initial years of carbon trading.

This chapter sheds new light on this issue by empirically analyzing the trading behavior of industrial firms on the EU carbon market. The study is based on a unique set of data which we collected between August and October 2009 by interviewing managers of approximately 800 manufacturing firms, 429 of which were participating in the EU ETS, in six European countries. The interviews focused on issues related to climate change and energy use and were conducted by a novel method that circumvents various types of bias that plague more traditional survey formats. Using data from the interviews, we obtain a number of new descriptive results regarding the behavior of regulated firms during the second trading phase of the EU ETS. Our first finding is that most installations bank permits from one year to the next. Second, about 30 percent of the firms do not appreciate the fact that the EU ETS has created a market for pollution—that is, they do not consider carbon allowances to be an asset that provides profit opportunities. Rather, they see the cap implicit in their allowance allotment as something they merely need to comply with. Although there are significant differences in EU ETS engagement between sectors, there are no significant differences between countries. Third, a large majority of plants operated by firms with more than one installation in the EU ETS manage their allocations of permits independently, even though they are allowed to pool permits across installations for more efficient abatement. Fourth, the majority (54 percent) of the 429 EU ETS participants interviewed does not trade on the EU allowance market, regardless of whether they consider the EU ETS more as a compliance mechanism or a cap-and-trade scheme. Some of these firms do not have to buy emissions allowances, because their emissions do not exceed the amount of allowances they were allocated. However, we also find evidence supporting the concern that firms do not make their excess allowances available on the market even when they should. On average, firms begin to sell allowances only if they have an excess supply of 5,000 to 10,000 allowances. Still, the total number of excess allowances held by firms below this "trading threshold" is less than 10 percent of all excess allowances.

9.1 Interviewing managers

Interview methodology

Our survey builds upon and substantially extends work on climate-change policies and management practices by Martin, Muûls, de Preux, and Wagner (2012). We conducted structured telephone interviews with managers at randomly selected manufacturing facilities in Belgium, France, Germany, Hungary, Poland, and the United Kingdom. The interview setup follows the management survey design pioneered by Bloom and van Reenen (2007, 2010) in that the interviewer engages interviewees in a dialogue with open questions that are meant not to be answered "Yes" or "No." On the basis of this dialogue, the interviewer then assesses and ranks the company along various dimen-sions. Interviewees do not know that the interviewers are scoring their answers; interviewers do not know the performance characteristics of the firm whose representative they are interviewing.

The interview format is designed to avoid several sources of bias that are common in conventional surveys (Bertrand and Mullainathan 2001). For instance, experimental evidence shows that a respondent's answers can be manipulated by making simple changes in the ordering of questions, in how questions are framed, or in the numerical scale on which respondents are supposed to score their answers. By asking open-ended questions and by delegating the task of scoring the answers to the interviewer, we seek to minimize cognitive bias. Cognitive bias on the part of the interviewers can be controlled for by using interviewer-fixed effects in the regression analyses. Another common observation about survey data is that respondents are tempted to report attitudes or patterns of behavior that are socially desirable but which may not correspond to what they actually think and do. This problem may be exacerbated in situations where respondents do not have a definite attitude toward the issues they are asked about but are reluc-tant to admit that. Our research design addresses this issue in two ways. First, the interviewer begins by asking an open question about an issue and then follows up with more specific questions, or asks for some examples in order to evaluate the respondent's answer as precisely as is possible. Second, the results of the interviews are then, as a validation exercise, linked to independent data on economic performance.

Interview practice
Using the Orbis database maintained by Bureau Van Dijk, we obtained contact details for 44,605 manufacturing firms in Belgium, France, Germany, Hungary, Poland, and the United Kingdom.[3] To solicit inter-viewees, we randomly selected companies from that list. To ensure sufficient coverage of EU ETS firms, we also sampled manufacturing firms at random from the CITL in these countries. Interviewers made "cold calls" to production facilities (not head offices), giving their names and their affiliation before asking to be put through to the envi-ronmental manager. In the case of EU ETS firms, interviewers asked for the person responsible for the EU ETS, who is listed in the CITL. Table 9.1 reports the number of calls made and various statistics about the response rates.

Table 9.1
Interview response rates by country.

	Number of interviews	Number of firms interviewed	Number of ETS firms interviewed	Number of non-ETS firms interviewed	Number of firms contacted	Number of firms that refused	Response rate
Belgium	134	131	85	46	178	47	0.74
France	141	140	92	48	238	98	0.59
Germany	139	138	95	43	337	199	0.41
Hungary	69	69	37	32	90	21	0.77
Poland	78	78	57	21	140	62	0.56
UK	209	205	63	142	468	264	0.44
Total	770	761	429	332	1451	691	0.52

Note: There are more interviews than interviewed firms because we conducted several interviews with different partners in a small number of firms.

An ordinal scale of 1 to 5 was used to measure various management practices related to climate change. For each aspect of management ranked in this way, interviewers were instructed to ask a number of open questions. Questions were ordered so that the interviewer began with a fairly open question about a topic, then (if it seemed necessary) probed for more details in subsequent questions. The goal was to gauge the practices of firms according to a few common criteria. For instance, rather than ask for a subjective assessment of the extent to which a firm's management was aware of climate-change issues, we gauged this by how formal and far-reaching the management's discussion of climate-change topics was. To verify the consistency of the interviewer's scoring, a subset of randomly selected interviews was double-scored by a second team member who listened in. The purpose of the interviews was to gather information on both the effectiveness and the competitiveness effects of climate-change policies, particularly those of the EU ETS, in a random sample of European manufacturing firms.

The questionnaire is divided into broad sections dedicated to (i) the current and anticipated future effects of the EU ETS, (ii) prices for energy and CO_2, competition and other external drivers of climate-change related management practices, and (iii) specific measures that were adopted by firms and other measures that were considered but eventually discarded. Additional information on relevant company characteristics is also collected.

9.2 Descriptive evidence of behavior in the allowance market

This section provides descriptive evidence of the behavior of EU ETS firms in the allowance market, using the responses to a series of interview questions pertinent to this issue. Economists often assume that firms make decisions according to the price of emissions and act in rational profit maximizing ways. However, as Brewer (2005) and McKinsey and Ecofys (2006) point out, there was much uncertainty in the business community about the implementation of the EU ETS, about its functionality, and about its credibility as a tool for reducing greenhouse-gas emissions. Both studies put forward the view that the "market rationality" of EU ETS participants probably was limited due to the confusion over the design of current policy and the stringency of future policy. For these reasons, rather than take rationality for granted, we analyze the responses to interview questions that help us understand how firms act in the market and how they make decisions.

Banking

The interviews were conducted during Phase II of the EU ETS (2008–2012). In Phase II, market participants were allowed to either borrow allowances from their allocation of the following year or bank excess allowances for future use. By the last day of March each year, the operator had to submit the verified annual emissions to the EU ETS Registry, endorsed by a verifier. By April 30, a sufficient number of allowances to match the installation's emissions had to be surrendered. Because operators received their allowance allocation for the following year already at the beginning of April, they could "borrow" those permits when surrendering permits for the previous year's emissions. We first asked managers about their position at the end of March:

[Q1] In March of this year (i.e., 2009, before the compliance cycle), what was your allowance position on this site? Were you short or long in allowances?

As figure 9.1 shows, a large majority of installations had an excess of permits to cover their emissions. Both the generous allocation process and the economic downturn could explain such a pattern.

During the month of March, firms are allowed to sell their excess permits. Firms that are short can either buy or borrow permits to be in

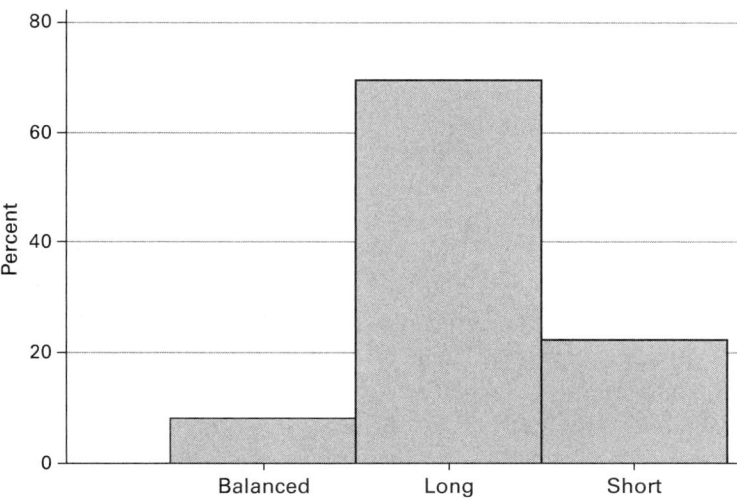

Figure 9.1
Allowance positions of EU ETS installations in March 2009, before the compliance period.

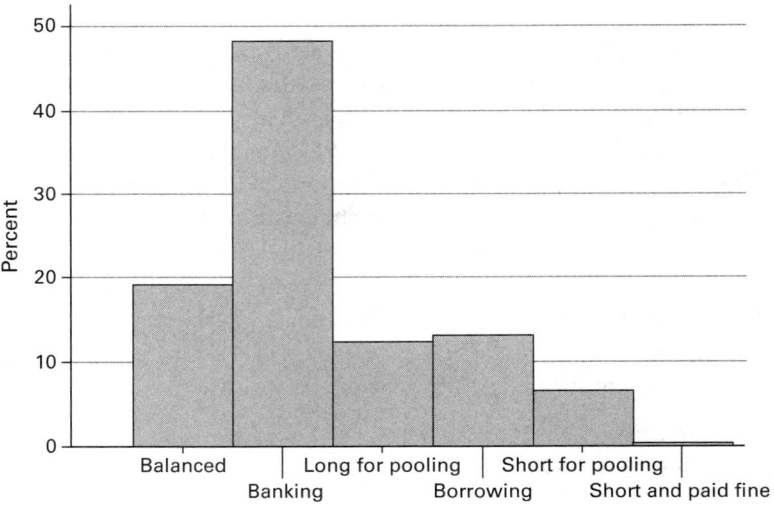

Figure 9.2
Allowance positions of EU ETS installations in April 2009, after the compliance period

compliance. With regard to this compliance month, the interviewees were asked the following:

[Q2] In April this year, what was your position after the compliance process?

{If answers "long" } Did you bank permits for future years? Why?

{If answers "balanced/compliant" or "short" } Did you borrow permits from next year's allowance? Why?

{If answers "short" } Why did you remain short?"

Figure 9.2 summarizes the answers to these questions.[4] Not more than 20 percent of firms were in a balanced position after this clearing period. In contrast, most firms that had excess permits before the compliance chose not to sell them and banked them for future years. Because the price in March and April of 2009 was lower than €10, this suggests that firms anticipated an increase in the price in future years and perhaps a decline in their allocation of free permits in Phase III. In most sectors except Ceramics, a large majority of firms were either balanced or long in allocations after the compliance period. (See figure 9.3.)

Figure 9.2 shows that nearly 20 percent of interviewed installations reported being either long or short in allowances after the compliance

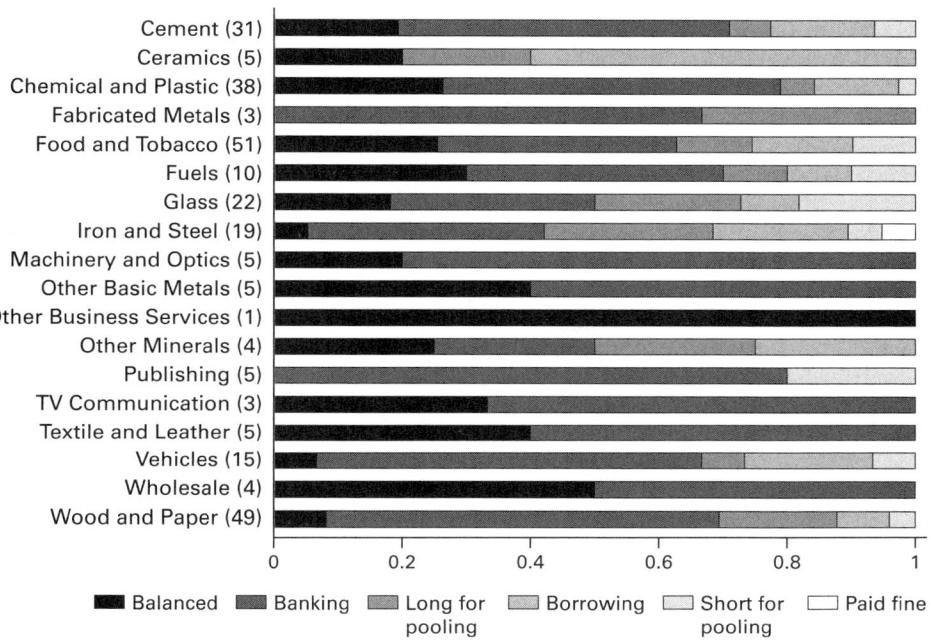

Figure 9.3
Allocation positions of EU ETS installations by sector in April 2009, after the compliance period.

period for "pooling reasons." The EU ETS design in Phase II also allowed installations to pool their allowances across installations within the same company. By far the most common reason for remaining short after the compliance period is that the installation is part of a larger group of EU ETS plants that are pooling their allowances. Only one installation reported being fined for having stayed short after compliance.

Managers of multi-plant firms were also asked where their installation's compliance was managed:

[Q3] Is EU ETS compliance managed on the production site or elsewhere?

Figure 9.4 summarizes the answers to this question. It is striking that a large majority of firms manage their allocations of permits within their own installation—even when the firm has more than one EU ETS installation.

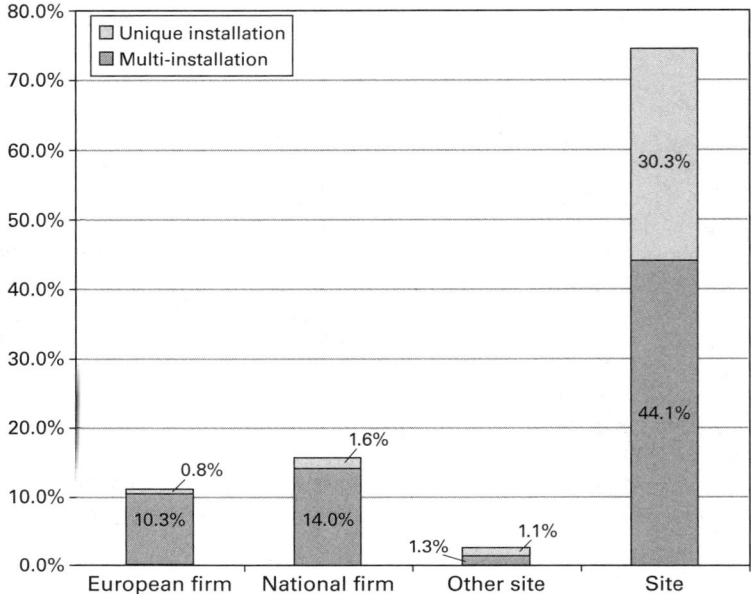

Figure 9.4
Centralization of trading decisions as indicated by where compliance with the EU ETS is managed. Unique installations are those that are the only plants of their firm or group in the ∃U ETS.

Raticnality

To m∋asure whether the trading system had been fully understood as an economic tool that could be used to minimize compliance costs, we asked specific (closed-ended) questions about trading behavior on the perm it market pertaining to the frequency, net balance, banking, and borrowing of European Union Allowances; we also asked an open-ended question about how the firm made trading decisions and assessed the response against a set of showcase behaviors characterized by an increasing degree of sophistication. Question 4 was as follows:

[Q4] How do you decide how many permits to buy or sell or trade at all? Did you base this decision on any forecast about prices and/or enerзy usage? Did you trade permit revenue off against emission reduction costs in your planning on this issue?

Responses were scored on an integer scale ranging from 1 to 5. A low "rationality score" would be recorded if the firm took its permit alloca-tion as a target to be met and did not take into account the price of permits or the cost of abatement and would sell if there was a surplus or buy if there was a deficit. A high score would correspond to a company with a thorough understanding of the site-specific CO_2 abatement cost using trading both to reduce compliance cost and to generate extra revenues from excess abatement. To achieve a very high score, a company must form expectations about permit prices, re-optimize abatement choices if that is necessary, and use futures and derivatives to manage EUAs as a financial asset.

The results broadly support the findings of Brewer (2005) and McKinsey and Ecofys (2006). The average rationality score of 2.56 (out of 5) suggests that the average firm does not follow an overly rational strategy when trading EUAs. According to the scoring criteria, a score of 3 means that a firm is in the process of learning how the market works. (Firms often take a permit allocation as a target to be met and fail to factor the price of permits or the cost of abatement into management decisions.)

Figure 9.5 depicts the distribution of the rationality score, which captures a firm's vision of trading opportunities in the EU ETS. About 30 percent of firms participate only passively in the EU allowance

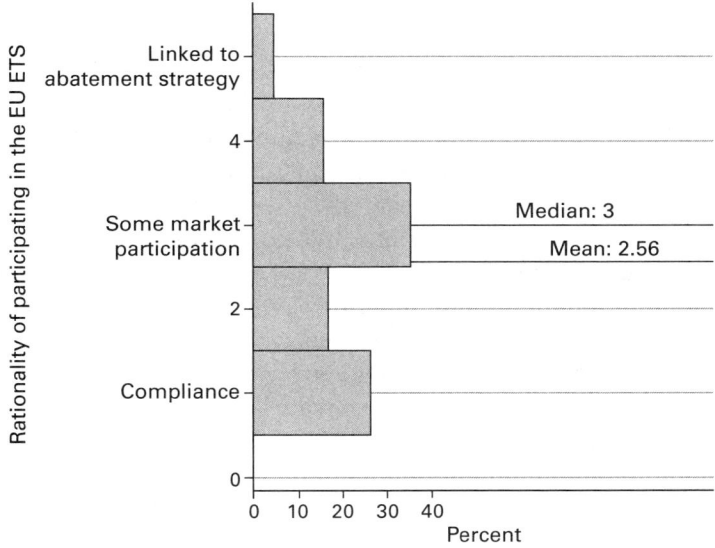

Figure 9.5
Rationality of market participation as indicated by the interview score measuring whether the firm is acting rationally on the EU ETS market.

market, seeing the EU ETS as something they merely need to comply with. These firms take their permit allocation as a target to be met, much in the spirit of "command-and-control" regulation. This prevents a firm from minimizing its abatement cost, which, on theoretical grounds, makes permit trading superior to command-and-control regulation. For example, the total compliance cost of a given emission cap will not be minimized unless firms rationally choose abatement levels such that their marginal abatement cost equals the permit price.

Although there are significant differences in ETS engagement between sectors, there are no statistically significant differences between countries. (See figure 9.6a.) In contrast, figure 9.6b shows that some industries (i.e., cement, chemicals and plastics, fuels) are seizing permit market opportunities more efficiently than others.

Trading

Managers were asked specifically about their trading behavior:

[Q5] Eefore the compliance process in April, did you buy or sell allowances on the market or over the counter from other firms? If not, why not? If yes, how frequently?

Figure 9.7 shows that 54 percent of EU ETS participants do not trade on the EU ETS market. We study this further by looking at the frequency with which firms trade. Figure 9.8 shows that firms trade permits at least quarterly in five of the six sample countries across a variety of industries.[5] The first column of table 9.2 reports the result of an ordinary least-squares regression of the rationality score on the frequency of trading, with the emissions, the employment, the sector, and the country of the firm controlled for. It shows that firms that are more rational trade more frequently. Some firms need not trade, because their emissions exactly match their allocated allowances; others may have excess allowances that they do not supply to the market. Some policy makers are concerned that this behavior exacerbates the shortage of allowances and hence drives up the allowance price. There may be different reasons for failure to sell excess allowances. For example, a firm may want to bank permits in order to hedge against future increases in the carbon price, or a firm may face high transaction and information costs related to trading, or a firm may fail to optimize. Murphy and Stranlund (2007) suggest that an "endowment effect"—overvaluation of items in one's possession—may prevent firms from

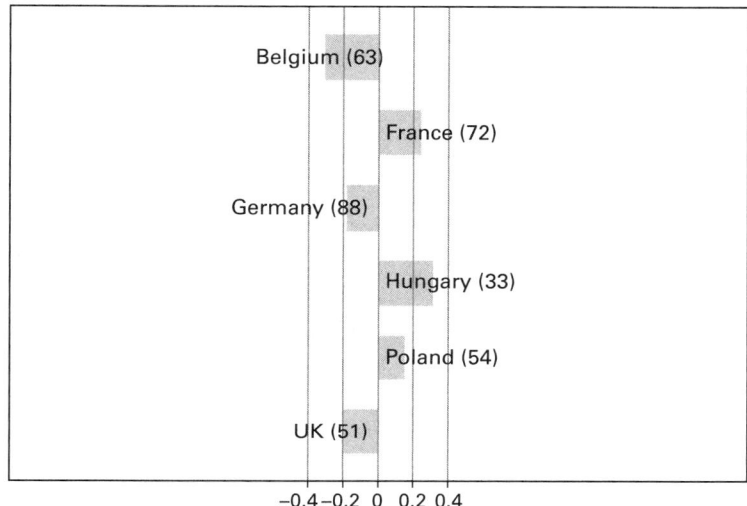

(a)   Less rational than others   More rational than others

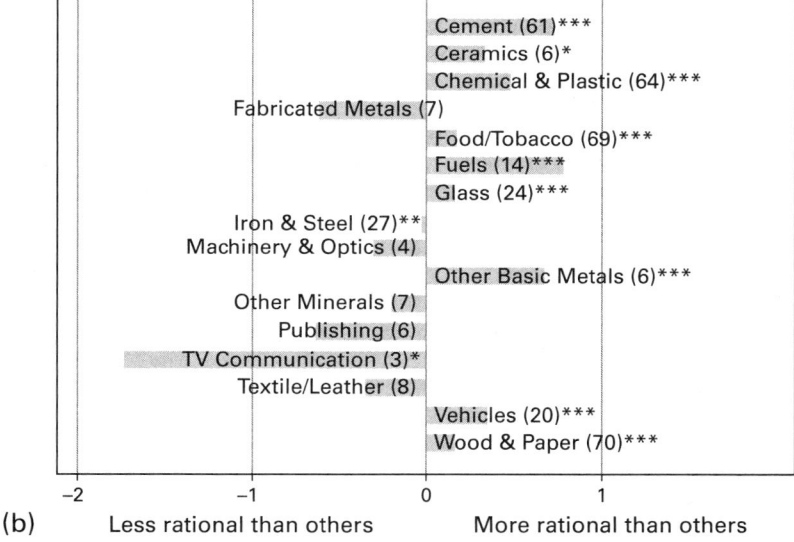

(b)   Less rational than others   More rational than others

Figure 9.6
Rationality of market participation across countries and sectors (a) across countries with three-digit sectors controlled for and (b) across sectors. Each graph shows the average difference—conditional on noise controls—between firms from different countries or sectors in terms of the interview score for market rationality depicted in figure 9.5. Three asterisks indicate statistical significance level of 1 percent; two asterisks indicate 5 percent; one asterisk indicates 10 percent.

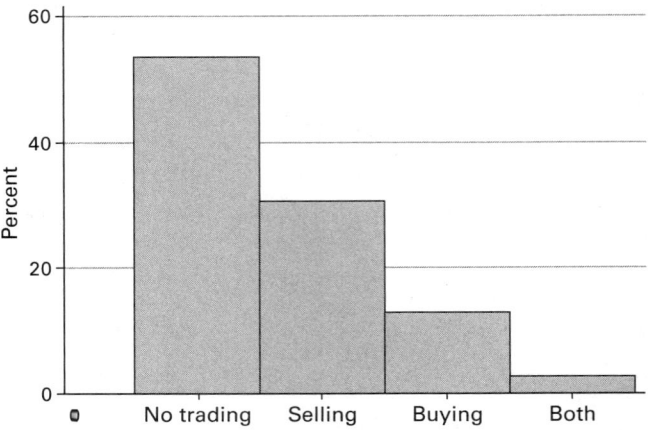

Figure 9.7
Participation in the EU ETS market.

selling permits they were allocated for free. A "status quo bias" (Kahneman et al. 1991; Samuelson and Zeckhauser 1988) can have similar consequences.

9.3 Regression analysis

In this section we analyze the hypothesis that firms fail to sell excess allowances. We run probit regressions on the binary event "selling on the EU ETS market" derived from question Q5. The main explanatory variable includes a set of dummy variables that characterize the distribution of excess allowances across firms.

We hence define a variable $EXCESS_i$ as

$$EXCESS_i = \max\{ALLO_i - CO2_i, 0\},$$

where $ALLO_i$ and $CO2_i$ are, respectively, the permit allocation and the CO_2 emissions of firm i. The variables correspond to figures for the year 2008 taken from the CITL.[6] We split firms with positive excess allowances into ten equally sized groups according to the size of their excess allocation. In other words, the ten groups are defined by the deciles of the distribution of excess allowances (960, 1,685, 2,888, ... , 27,405, 55,948 allowances). If we let each group be represented by a dummy variable $Q_{d,i}$, where the subscript d denotes the decile, we can express the latent equation underlying the Probit as

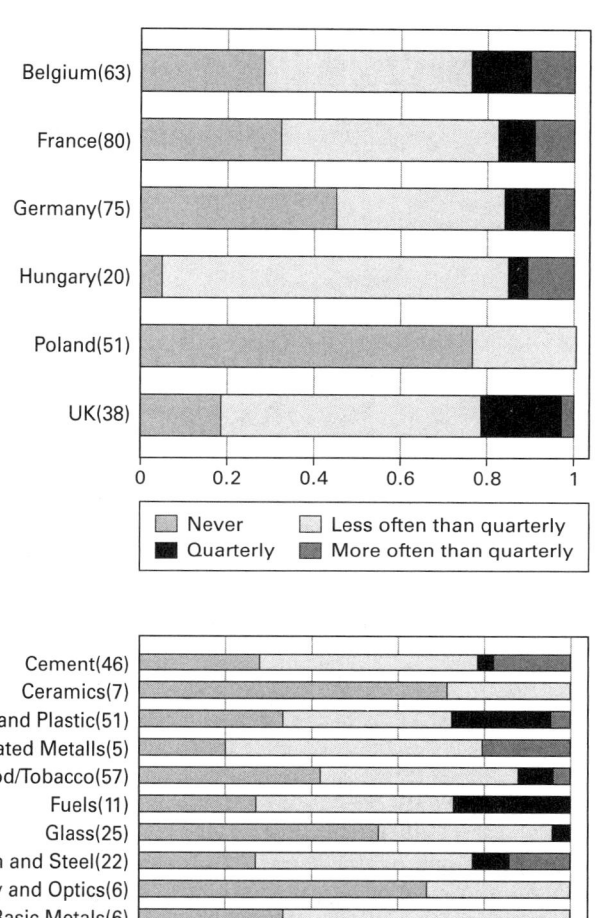

Figure 9.8
Frequency of permit trading.

Table 9.2
Correlation between rationality and frequency of trading or excess allocation.

Dependent variable	EU ETS rationality score	
	(1)	(2)
Frequency of trading	0.252***	
	(0.081)	
Over-allocation		0.193***
ln(allocation − CO₂)		(0.057)
CO₂ consumption	0.076	−0.026
ln(CO₂)	(0.046)	(0.057)
Employment	0.107*	0.096
	(0.057)	(0.064)
Sector and country controls?	Yes	Yes
Observations	315	249

This table presents the results of OLS regressions with the dependent variable derived from interview responses to question 4. Frequency of trading is the score variable derived from responses to question 5. CO_2 consumption and over-allocation in 2008 data are taken from the CITL dataset. Employment and Sector are obtained from Orbis. Robust standard errors are reported in parenthesis. Stars indicate statistical significance level: ***=1 percent, **=5 percent, *=10 percent.

$$SELL^*_i = \sum_d \beta_d Q_{d,i} + \beta_x X_i + \varepsilon_i \, ,$$

where X_i is a vector of additional control variables.

Table 9.3 reports the results of this and other specifications. Columns 1 and 2 display regressions of the event "trading on the ETS," which also includes buying in addition to selling. Trading is weakly correlated with the total amount of CO_2 a firm emits (column 1) but much more strongly correlated with the rationality score we derived from question Q4. This corroborates the internal consistency of this score. Columns 4–6 report specifications where the decision to sell allowances on the EU ETS market is the dependent variable. When dummies for the deciles of *EXCESS* are included, columns 5 and 6 show that only firms with excess allowance amounts in the fifth decile and higher have a strongly significant heightened probability of selling allowances on the EU ETS market.

Because table 9.3 reports marginal effects, the coefficient estimate has the following interpretation: A firm with about 4,700 allowances or more to spare has, on average, a 50 percent higher probability of selling some or all of those allowances on the EU ETS market. This implies that allowances are less likely to be sold on the allowance market when the revenue derived by their owners is small—a finding that could be rationalized by a fixed cost of trading. Column 3 reports a similar specification, but in this case the dependent variable is selling *or buying* in the market rather than selling only. It appears that threshold effects do not occur in the case of buying decisions.

An alternative specification is presented in the first column of table 9.4, where the excess of allowances is calculated relative to CO_2 emissions. Firms in the first decile also have a larger probability of selling allowances. This shows that it is the absolute and not the relative number of excess permits that is correlated with selling choices. Controlling for employment of the firm in column 2 does not affect the results either: larger firms are not more likely to be selling.

As figure 9.3 shows, a large share of responding market participants report that they pool their emissions, adapting their trading behavior accordingly. To test whether this might affect the selling patterns observed in table 9.3, we include in column 3 of table 9.4 a dummy indicating that the firm reported pooling its permits. Including this dummy has no effect on the results.

It is interesting to consider whether firms are holding on to these excess permits because of transaction costs or whether they are banking permits for future periods (perhaps because they expect the price to rise). This hypothesis is tested in column 4, where conditioning on the fact that the firm reported banking permits does not affect the results. Column 5 of figure 9.3 confirms that our results are robust to including sector and country controls.

How important is the withholding of excess allowances on aggregate? To answer this question, we examine what share of the excess allowances is held by firms in deciles 1–4. Figure 9.9 illustrates this share in the distribution of excess allowances. On aggregate, this problem seems to be of only minor importance, as far less than 10 percent of excess allowances fall into the "no trade" category.

Conclusion

A descriptive analysis of firms' allowance positions and trading patterns reveals that most installations across sectors were banking allowances at the end of the compliance cycle. Also, the permit position is managed primarily within individual installations. We also find that about 30 percent of firms participate only passively in the EU ETS; that is, they do not consider carbon allowances as a financial asset that

Table 9.3
Regressions of firms' trading decisions.

Dependent variable		(1)	(2)	(3)	(4)	(5)	(6)
		Buys or sells in ETS			Sells in ETS		
CO$_2$ consumption		0.033*	0.025	0.004	−0.007	−0.028	−0.033
		(0.018)	(0.019)	(0.022)	(0.017)	(0.022)	(0.022)
ln(CO$_2$)							
EU ETS Rationality Score			0.070***	0.065**	0.069***		0.054**
			(0.026)	(0.026)	(0.024)		(0.025)
Over-allocation in 2008 [number of permits]	0–960			−0.115		0.132	0.136
				(0.129)		(0.147)	(0.149)
	960–1685			−0.068		0.263*	0.258*
				(0.130)		(0.139)	(0.141)
	1685–2888			0.022		0.121	0.106
				(0.143)		(0.162)	(0.162)
	2888–4754			−0.210*		0.123	0.100
				(0.119)		(0.144)	(0.145)

	(1)	(2)	(3)	(4)	(5)
4754–6671			0.106	0.425***	0.426***
			(0.115)	(0.109)	(0.110)
6671–10870			0.117	0.500***	0.482***
			(0.115)	(0.096)	(0.101)
10870–15963			0.163	0.524***	0.503***
			(0.108)	(0.091)	(0.096)
15963–27405			0.120	0.516***	0.492***
			(0.123)	(0.099)	(0.104)
27405–55948			0.194*	0.540***	0.544***
			(0.109)	(0.091)	(0.091)
>55948			0.132	0.471***	0.450***
			(0.131)	(0.117)	(0.123)
Observations	299	299	299	299	299

This table presents the results of five probit regressions. The dependent variable is derived from interview responses to question 7. In columns 1 to 3 it takes value 1 if the firm is trading on the EU ETS. In columns 4 to 6, the variable only takes value 1 if the firm is only selling allowances. CO_2 consumption and over-allocation in 2008 data are taken from the CITL dataset. Stars indicate statistical significance level: ***=1 percent, **=5 percent, *=10 percent.

Table 9.4
Regressions of firms' trading decisions.

Dependent variable		(1)	(2)	(3)	(4)	(5)
		Sells in ETS				
CO₂ consumption						
ln(CO₂)		0.029	−0.029	−0.035	−0.030	−0.010
		(0.029)	(0.022)	(0.022)	(0.022)	(0.025)
EU ETS rationality score		0.052**	0.056**	0.054**	0.053**	0.075***
		(0.026)	(0.025)	(0.025)	(0.025)	(0.029)
Over-allocation in 2008 [number of permits]	0–960	0.292**	0.161	0.162	0.141	0.252
		(0.133)	(0.152)	(0.153)	(0.149)	(0.167)
	960–1685	0.161	0.260*	0.275*	0.249*	0.204
		(0.137)	(0.140)	(0.141)	(0.142)	(0.162)
	1685–2888	0.217	0.118	0.119	0.124	0.132
		(0.133)	(0.163)	(0.164)	(0.164)	(0.192)
	2888–4754	0.243*	0.112	0.119	0.107	0.016
		(0.139)	(0.146)	(0.148)	(0.146)	(0.156)
	4754–6671	0.356***	0.427***	0.458***	0.420***	0.410***
		(0.126)	(0.110)	(0.112)	(0.112)	(0.134)
	6671–10870	0.460***	0.479***	0.507***	0.477***	0.549***
		(0.107)	(0.102)	(0.101)	(0.103)	(0.109)

	(1)	(2)	(3)	(4)	(5)
10870–15963	0.325**	0.506***	0.526***	0.502***	0.444***
	(0.130)	(0.095)	(0.095)	(0.097)	(0.123)
15963–27405	0.628***	0.498***	0.508***	0.515***	0.464***
	(0.073)	(0.104)	(0.103)	(0.102)	(0.129)
27405–55948	0.471***	0.539***	0.565***	0.544***	0.557***
	(0.108)	(0.092)	(0.089)	(0.091)	(0.106)
>55948	0.660***	0.458***	0.472***	0.453***	0.521***
	(0.065)	(0.122)	(0.121)	(0.123)	(0.132)
Employment		−0.022			−0.004
		(0.023)			(0.029)
Banking			−0.052		
			(0.062)		
Pooling				−0.099	
				(0.072)	
Over-allocation relative to emissions	Yes	No	No	No	No
Sector and country controls	No	No	No	No	Yes
Observations	299	299	299	299	299

This table presents the results of five probit regressions. The dependent variable is derived from interview responses to question 7. It only takes value 1 if the firm is only selling allowances. CO_2 consumption and over-allocation in 2008 data are taken from the CITL dataset. Other variables are derived from interview responses. Stars indicate statistical significance level: ***=1 percent, **=5 percent, *=10 percent.

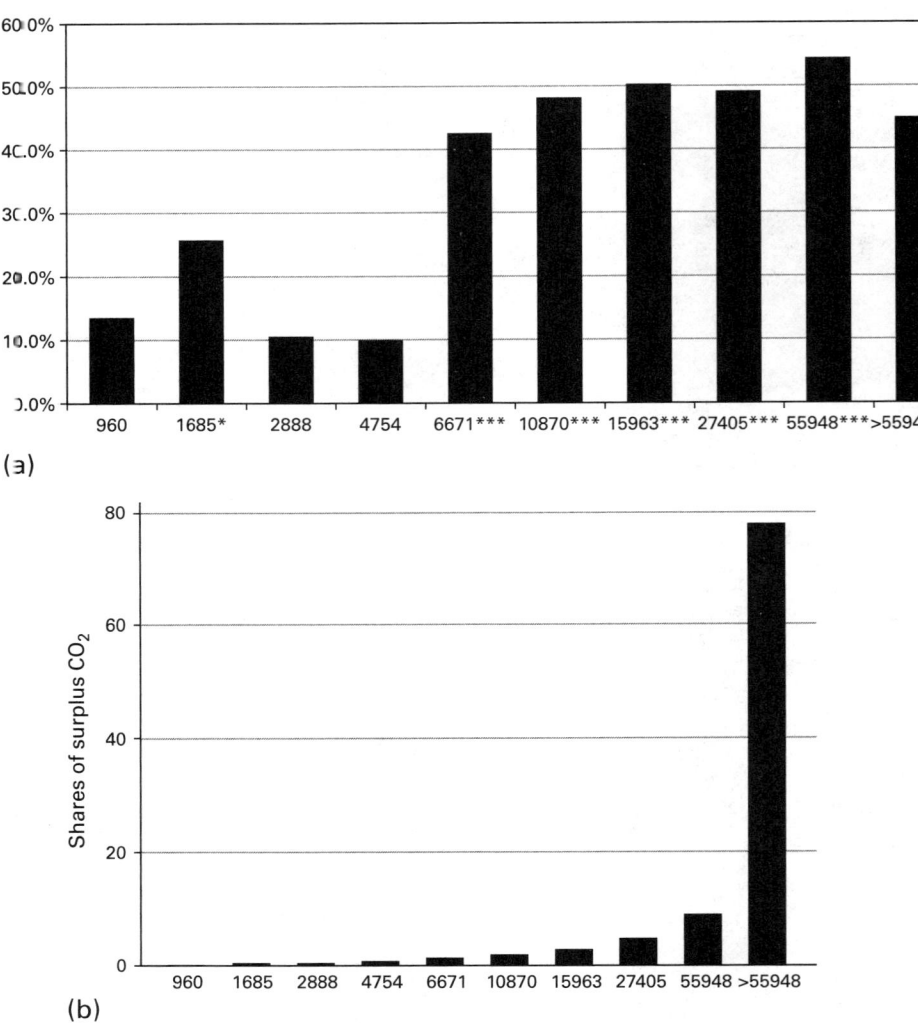

(a)

(b)

Figure 9.9
Probability of selling and distribution of excess allowances. Panel a presents the coefficients of the probit regression of column 4 of table 9.3, each bar representing the probability that a firm with excess allowances within that range will sell allowances on the EU ETS market. Panel b represents the shares of the total amount of excess allowances in each quintile. Three asterisks indicate statistical significance level of 1 percent; two asterisks indicate 5 percent; one asterisk indicates 10 percent.

could provide profit opportunities. Rather, they see the EU ETS as providing a cap on emissions with which they must comply. Though there are significant differences in EU ETS engagement between sectors, no significant differences arise between countries. In line with this observation, the majority of EU ETS participants in our sample does not trade on the EU allowance market. Some of these firms do not need to trade, because their emissions correspond to their allowance allocations. There is, however, concern that some firms do not make their allowances available even though they possess an excess supply. Our regression analysis provides empirical support for this. On average, firms begin to sell only if they have an excess supply of around 5,000 allowances. The total number of excess allowances held by firms below the trading threshold is rather small—less than 10 percent of all excess allowances. This is robust to controlling for country and sector fixed effects, for the size of the firm, and for the possibility that the firm is pooling or banking allowances.

In sum, our findings show that industrial firms engaged in the EU ETS in Phase II did not exploit the economic opportunities of the permit market to the fullest. However, it is unlikely that some firms holding on to their excess allowances had a large effect on the carbon price because of their rather small market share. In Phase III, the increased use of permit auctioning is likely to change the behavior of firms in the market, as they will increasingly have to bid for allowances in auctions or buy them from other market participants.

An important lesson arising from our findings is that the implementation of fully efficient carbon markets is still hampered by obstacles to trading that affect small participants in the market disproportionately, such as transaction costs or information barriers. Policy makers can mitigate these obstacles in various ways. To overcome information barriers, the regulator can provide free consulting services to market participants to make sure they are aware of the gains from trading permits across firms as well as of the benefits of pooling permits across installations within the firm. This would also lower transaction costs arising because of indivisibilities in trading services that some firms (presumably small ones) are likely to procure from external advisors and consulting firms. More generally, policy makers should take into account that, whenever a new carbon market is implemented, participants go through a phase during which they learn how to make the most of the system's efficiency potential.

Appendix: A parametric approach

In section 9.4 above we examined the relationship between excess permits and trading using a non-parametric approach by including a number of dummies for different levels of excess permits. Here we look at the same issue using a parametric approach. Specifically, we fit a quadratic function to the positive branch of the excess allowance distribution by estimating the regression

$$SELL^*_i = \beta_0 (EXCESS_i \leq 0)$$

$$+ (EXCESS_i > 0)\big[\beta_1 \ln(EXCESS_i) + \beta_2 \ln(EXCESS_i)^2\big] + \beta_x X_i + \varepsilon_i \,.$$

We show the result of this specification by plotting the relationship between excess permits and the probability of trading in figure 9.10. The trading probability is significantly different from zero at a level of excess permits above 500. This is somewhat lower than what we found with the non-parametric approach in the main text. However, qualitatively the result is the same in that there seems to be positive level of excess permits below which trading does not seem to occur.

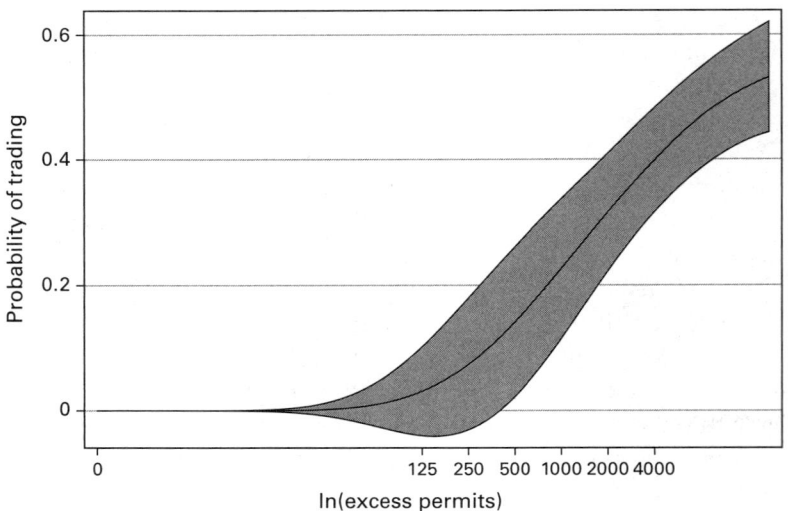

Figure 9.10
Propensity to trade allowances and excess allocation as indicated by results of using a parametric approach to analyze the relationship between excess permits and trading. The scale of the horizontal axis is logarithmic.

Notes

1. Independence also requires competitive permit markets and lack of substantial transaction costs and regulatory uncertainty.

2. The CITL has recently changed its name to European Union Transaction Log (EUTL). In the remainder of the chapter, we use the old name.

3. For more details on the survey, see Martin et al. 2014.

4. It is interesting to compare these answers to data available in the CITL on allocations, surrendered permits, and verified emissions for 2009. In 74 percent of cases these correspond perfectly—for example, the manager of a firm that has more allocated permits than verified emissions is reporting that his firm is banking. However, it could be the case that the firm bought or sold allowances and was in a different position after the compliance process, and this information is not available for 2009.

5. Anecdotal evidence suggests that Polish firms failed to trade more frequently because of institutional barriers and delays, not because of a lack of aptitude or willingness to engage in the market. For instance, Skjærseth and Wettestad (2008, p. 281) note that "Poland must be counted as figuring centrally among the ETS implementation laggards so far. It was seriously delayed in NAP I."

6. Column 2 of table 9.2 shows that excess allocation is positively correlated to the rationality score described above. This suggests that firms that receive more EUAs than they need to cover their emissions behave more rationally in the market than other firms.

References

Bertrand, M., and S. Mullainathan. 2001. Do people mean what they say? Implications for subjective survey data. *American Economic Review* 91 (2): 67–72.

Bloom, N., and J. van Reenen. 2007. Measuring and explaining management practices across firms and countries. *Quarterly Journal of Economics* 122 (4): 1351–1406.

Bloom, N., and J. van Reenen. 2010. New approaches to surveying organizations. *American Economic Review* 100 (2): 105–109.

Brewer, T. L. 2005. Business perspectives on the EU emissions trading scheme. *Climate Policy* 5 (1): 137–144.

Ellerman, A. D., and B. K. Buchner. 2007. The European Union emissions trading scheme: Origins, allocation, and early results. *Review of Environmental Economics and Policy* 1 (1): 66–87.

Hahn, R. W., and R. N. Stavins. 2011. The effect of allowance allocations on cap-and-trade system performance. *Journal of Law & Economics* 54 (S4): S267–S294.

Jaraitė-Kažukauske, J., and A. Kažukauskas. 2014. Do transaction costs influence firm trading behaviour in the European Emissions Trading Scheme? *Environmental and Resource Economics*, in press.

Kahneman, D., J. L. Knetsch, and R. H. Thaler. 1991. Anomalies: The endowment effect, loss aversion, and status quo bias. *Journal of Economic Perspectives* 5 (1): 193–206.

Löschel, A., P. Heindl, V. Alexeeva-Talebi, V. Lo, and A. Detken. 2010. KfW/ZEW CO 2 Panel: Vermeiden oder kaufen—Deutsche Unternehmen im Emissionshandel. *Zeitschrift für Energiewirtschaft* 34 (1): 39–46.

Martin, R., M. Muûls, L. B. de Preux, and U. J. Wagner. 2012. Anatomy of a paradox: Management practices, organizational structure and energy efficiency. *Journal of Environmental Economics and Management* 63 (2): 208–223.

Martin, R., M. Muûls, L. B. de Preux, and U. J. Wagner. 2014. Industry compensation under relocation risk: A firm-level analysis of the EU Emissions Trading Scheme. *American Economic Review* 104 (8): 2482–2508.

McKinsey and Ecofys. 2006. Review of Emissions Trading Scheme: Survey highlights. Technical report, European Commission Directorate General for Environment, Brussels.

Montgomery, W. 1972. Markets in licenses and efficient pollution control programs. *Journal of Economic Theory* 5 (3): 395–418.

Murphy, J. J., and J. K. Stranlund. 2007. A laboratory investigation of compliance behavior under tradable emissions rights: Implications for targeted enforcement. *Journal of Environmental Economics and Management* 53 (2): 196–212.

Pinkse, J., and A. Kolk. 2007. Multinational corporations and emissions trading: Strategic responses to new institutional constraints. *European Management Journal* 25 (6): 441–452.

Samuelson, W., and R. Zeckhauser. 1988. Status quo bias in decision making. *Journal of Risk and Uncertainty* 1 (1): 7–59.

Sandoff, A., and G. Schaad. 2009. Does EU ETS lead to emission reductions through trade? The case of the Swedish emissions trading sector participants. *Energy Policy* 37 (10): 3967–3977.

Skjærseth, J., and J. Wettestad. 2008. Implementing EU emissions trading: success or failure? *International Environmental Agreement: Politics, Law and Economics* 8 (3): 275–290.

10 Tradable Emissions Permits with Offsets

Nathan Braun, Timothy Fitzgerald, and Jason Pearcy

This chapter extends the textbook model of tradable emissions permits to include two kinds of compliance assets: permits and offsets. Earlier applications of tradable emissions permits, such as SO_2 trading in the United States, did not include offsets. Another example is the European Union Emissions Trading System, currently the world's largest mandatory carbon trading scheme. During the first phase of the EU ETS (from 2005 to 2007), European Union Allowances (EUAs) were the only tradable asset available to emitters. Since 2008 the EU ETS has been linked to the Kyoto Protocol's flexible mechanisms, the Clean Development Mechanism (CDM) and Joint Implementation (JI), which generate alternative tradable compliance assets that offset emissions elsewhere in the world. Emitters are now able to use EUAs as well as Certified Emissions Reductions (CERs) and Emission Reductions Units (ERUs), which offset some of their verified emissions. All compliance assets, whether they are permits (EUAs) or offsets (CERs and ERUs), allow an installation to emit the equivalent of one metric ton of CO_2. With the prospect of additional future carbon trading schemes and linkage between existing schemes, the applicability of offsets is an important issue (Anger 2008; Klepper 2011).

The model presented in this chapter is applicable to any emissions-control system with multiple compliance assets, but observed outcomes from the EU ETS motivate our analysis. In particular, we observe that during Phase II of the EU ETS permit prices were greater than offset prices, and emitting installations within the EU ETS surrendered varying amounts of offsets. For the most basic model with permits and offsets, we show that neither of these empirical observations is consistent with the theoretical model. Our framework allows us to incorporate two extensions present in the EU ETS into the model: offset quotas and offset transaction costs. We also show that either offset quotas or

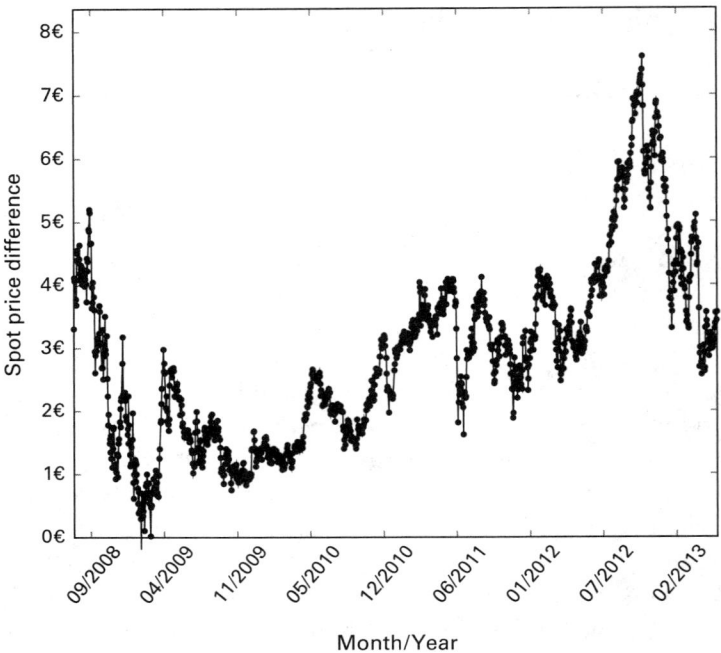

Figure 10.1
Spot price differential between permits (EUA) and offsets (sCER), August 2008–June 2013.
sources: IntercontinentalExchange, BlueNext

offset transaction costs explain why the price of a permit is greater than the price of an offset, but that offset quotas alone are not sufficient for explaining observed behavior within the EU ETS.

Two interesting observed outcomes from Phase II of the EU ETS are that permit prices are consistently above the price of offsets, and there is a large variance in the amount of offsets surrendered by an installation.[1] Figure 10.1 plots the average inter-day difference between the spot prices of permits and offsets over time.[2] Note that the price differential is always positive (except for February 11, 2009) and fluctuating over time. For 2010 and later years, the price differential remains above one euro, with a slight upward trend indicating that the price differential is not dissipating over time. Table 10.1 lists summary statistics for the amounts of offsets surrendered over time in the EU ETS.[3] The average number of offsets surrendered is increasing over time, as is the number of installations that surrender offsets. Note that the

Table 10.1
Offsets surrendered by installations. (Reported measures are determined conditional on an installation surrendering offsets that year.)

	Mean	S.D.	Maximum	Minimum	Number of installations
2008	40,461	160,727	4,984,978	1	1,630
2009	41,086	167,812	3,612,200	1	1,734
2010	45,690	187,048	4,716,145	1	2,674
2011	62,036	284,268	7,870,000	1	3,742
2012	78,847	314,814	7,503,306	1	5,752

source: EU Transaction Log (EUTL)

number of offsets surrendered varies widely across installations and that the variance is increasing.

Two results obtained with the basic model we develop in section 10.1 below are that the equilibrium price of a permit is equal to the equilibrium price of an offset and that all firms surrender the same amount of offsets.[4] In light of figure 10.1 and table 10.1, these theoretical predictions do not match up with the empirical observations. In section 10.2 and in later sections, we extend the basic model with offsets allowing for offset quotas and offset transaction costs as a means of making the model consistent with observed behavior.[5]

We consider offset quotas that were in place during Phase II for installations within the EU ETS. Quotas ensure that offsets supplement rather than replace emissions reductions. Table 10.7 below describes the offset restrictions that are in place for Phase II of the EU ETS. These restrictions are listed in each country's national allocation plan, and the offset quota for each installation is typically specified as a percentage of each installation's Phase II EUA (permit) allocation. Most countries allowed installations to bank offsets and to borrow against their annual offset quota in Phase II, and a minority of countries have special offset quotas that apply only to installations in certain sectors. Offset quotas are discussed in descriptive analyses by Trotignon (2012), Kettner, Kletzan-Slamanig, and Köppl (2012), Sterk and Wang-Helmreich (2008), and Lückge and Peterson (2004). Klepper and Peterson (2006) incorporate offset quotas into their CGE (computable general equilibrium) analysis to simulate a scenario with offset quotas and a scenario without them.

In this chapter, in addition to considering offset quotas, we incorporate offset transaction costs in our model. There is a growing literature on the implications of transaction costs in markets for tradable emissions permits and specifically within the context of the EU ETS. Discussion of transaction costs in such markets goes back to Stavins (1995). Michaelowa and Jotzo (2005) were the first to combine the theory behind offset transaction costs with an empirical evaluation, but they focused on the broader offset market and on transaction costs in credit creation. Specifically for the EU ETS, Jaraitė, Convery, and Di Maria (2010) and Heindl (2012b) use survey data to estimate transaction costs for Irish and German installations respectively during Phase I (2005–2007). Heindl (2012a) extends his previous analysis to consider how transaction costs and firm size affect the decision of German firms to either trade through an intermediary or directly on an exchange. Jaraitė and Kažukauskas (forthcoming) use Phase I transaction data to estimate how transaction costs influence an installation's decision to purchase permits and its decision as to how many permits to purchase.

In this chapter, we focus on transaction costs related to the surrendering of offsets. There are transaction costs associated with surrendering either permits or offsets. We posit that there are additional transaction costs associated with surrendering offsets. These additional costs might take place in the form of additional brokerage costs, informational costs associated with learning about the types of offsets and specific offset restrictions, or reputational costs associated with surrendering offsets. The approach we take is to normalize the transaction costs of permits to zero and to only consider transaction costs that pertain to offsets in excess of those for permits. With this approach, offset transaction costs are capturing the transaction costs associated with offsets that are in addition to the transaction costs associated with permits.

A large literature examines the characteristics of permit and offset prices within the EU ETS.[6] Some studies have examined factors that influence permit prices during Phase I (Alberola, Chevallier, and Cheze 2008; Hintermann 2010) and if these factors are still relevant during Phase II (Creti, Jouvet, and Mignon 2012); other studies have focused only on the prices of offsets (Conte and Kotchen 2010). A growing subset of the price literature examines the interactions between the prices of permit and the prices of offsets.[7] A majority of the studies examining the interactions between the prices of permits and those of offsets either are purely technical (for example, looking at co-integrating relationships between permit and offset prices) or also try to explain permit and offset prices by including some external price determinants.

Our model is also directly relevant to the literature focused on the linkage between different tradable permit systems, insofar as installations in the EU ETS can surrender offsets only if there is a linkage to the CDM and the JI. Market design issues arising in the linking of different tradable permit markets have been explored by a number of previous authors. Much of their work has focused specifically on the interaction between the EU ETS and the offset credits available through the CDM. Fankhauser and Hepburn (2010) make an observation directly applicable to the model presented below: that credits associated with baseline emissions inherently entail higher transaction costs. Michaelowa and Jotzo (2005) explore the magnitude of the transaction costs associated with creating the offsets. Wara (2007) identifies concerns about additionality of offsets as a key source of costs, citing the example of credits based on triflouromethane. More generally, Metcalf and Weisbach (2012) consider the inherent difficulties in linking heterogeneous systems created with different objectives in mind. Many tradable emissions permits schemes are intended to allow for cost-effective achievement of a specified goal. In contrast, baseline and offset schemes are likely to have physical as opposed to economic parameters, such as indexing pollution or emissions to a physical baseline. The difference in the goals of these two systems makes linking more costly.

Flachsland, Marschinski, and Edenhofer (2009) present a framework for evaluating linking tradable permit schemes. The primary linkage between the EU ETS and CDM offset credits that motivates the model we present below is a restriction on the number of credits that can be surrendered. However, we show that this particular linkage is neither necessary nor sufficient to explain observed behavior. Tuerk, Mehling, Flachsland, and Sterk (2009) explore the potential for linkages between different tradable permit systems through a third system (such as the CDM) and emphasize the role of the existing offset markets in establishing linkage norms. The connection between price-based and quantity-based regulatory schemes is explicitly considered by Wood, Heindl, Jotzo, and Löschel (2013).

The static model we develop in this chapter provides theoretical underpinnings for observed EU ETS outcomes as well as for the relationship between permit and offsets prices. We focus not only on prices, but also on market equilibria where prices, emissions, and surrendering decisions are endogenously determined. We incorporate institutional details specific to the EU ETS (offset quotas and offset transaction costs) into the model, then show how these details affect market equilibria. We then present descriptive statistics from the EU ETS to relate observed outcomes back to the theoretical implications derived from our model.

10.1 Model

Consider a static model in which firms abate and buy or sell permits with no transaction costs. Firms do not surrender offsets. In this model, the supply of permits and the demand for them are endogenously determined by the heterogeneity of abatement costs relative to permit endowment for installations covered under the emissions trading scheme.[8] With permits and offsets, the supply of permits is determined in the usual fashion, but the supply of offsets is exogenous and largely determined by external factors. In the EU ETS, a majority of the offsets surrendered originate in countries outside of the EU. By including offsets, the emissions trading scheme is no longer a closed system, and the linkage between permits and offsets allows for a broader array of compliance strategies.

There is an unit mass of firms, each firm having a marginal abatement cost curve MAC_i relating firm i's emission level e_i to its marginal abatement costs. Marginal abatement costs are increasing in abatement and decreasing in emissions. All regulated firms' marginal abatement cost curves are aggregated to form the market marginal abatement cost curve, MAC_M. With a unit mass of firms, we avoid any unnecessary aggregation issues and the market marginal abatement cost curve is also the marginal abatement cost curve of a representative firm. Consider two different subsets of firms: $\alpha \in (0, 1)$ firms with high marginal abatement costs (MAC_H) and $1 - \alpha$ firms with low marginal abatement costs (MAC_L).

As is typical in a tradable emissions permit system, the policy maker sets an emissions cap, $\bar{e}$, for the market and then creates a number of emissions permits that add up to the total cap on emissions. Permits are allocated to the firms and the market marginal abatement cost curve, MAC_M, describes the market demand for permits.[9] Figure 10.2 illustrates the different marginal abatement cost curves and an emissions cap of $\bar{e}$. In that figure, all firms have an unrestricted level of emissions of e^{max} at which marginal abatement costs are zero. As emissions decrease and abatement increases, the marginal abatement costs are increasing.

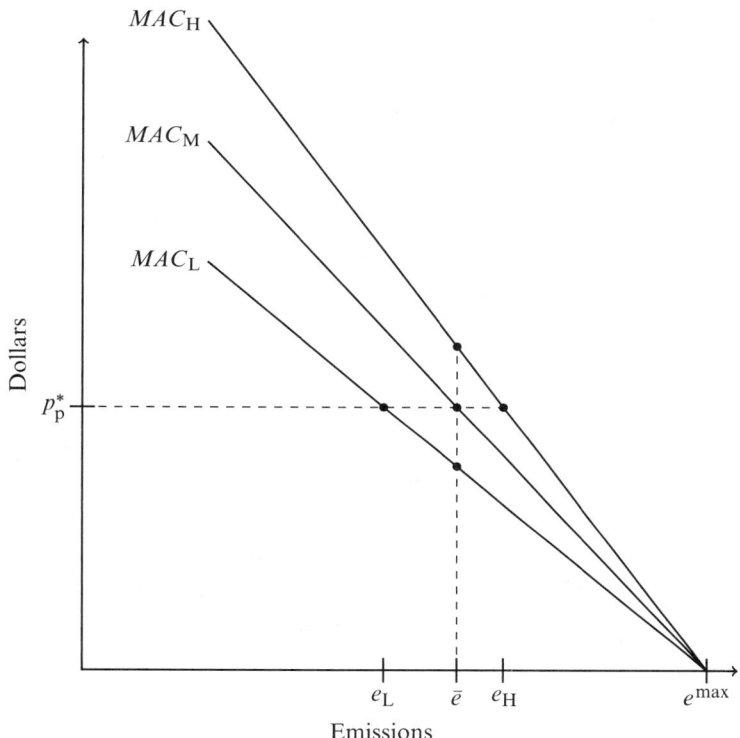

Figure 10.2
Static model with tradable permits and no offsets: heterogeneity in abatement costs creates gains to trade.

If each firm is initially given $\bar{e}$ permits, the total number of permits is $\alpha \bar{e} + (1 - \alpha)\bar{e} = \bar{e}$. In autarky, each type of firm emits $\bar{e}$ and the marginal abatement costs differ. With trade, firms exchange emissions permits due to the heterogeneity in marginal abatement costs.[10] Trading occurs until each firm's marginal abatement cost is equal to the equilibrium permit price,

$$p_{p}^{*} = MAC_{M}(\bar{e}) = MAC_{i}(e_{i}).$$

Allowing for costless trading, low marginal abatement cost firms emit e_{L} and high marginal abatement cost firms emit e_{H}. Firms with lower marginal abatement costs emit less than firms with higher marginal abatement costs and then sell permits to high marginal abatement cost firms. The total level of emissions is $\alpha e_{H} + (1 - \alpha)e_{L} = \bar{e}$. The volume of trade is influenced by the total number of permits issued, the composition of firm types, and the relative dispersion of firms' marginal abatement cost curves.

Unlike the supply and demand of emissions permits, the supply of offsets is assumed to be exogenous and determined by factors outside of the emissions-control system. Because, for example, a majority of offsets originate from projects in either China or India (Kettner et al. 2012; Trotignon 2012), one might assume that the firms within the ETS do not affect the supply of offsets. A non-decreasing (inverse) supply curve for offsets determines the relationship between the offset price, p_o and the quantity of offsets used, o, according to the equation $p_o = S(o)$.

First we allow emissions permits and offsets to be perfect substitutes, insofar as they entitle a firm to the same level of emissions.[11] The equilibrium determined when permits and offsets are perfect substitutes serves as a useful baseline. When offsets are incorporated into the tradable emissions permit system, firms have a larger set of compliance strategies. The equilibrium concept is the same with or without offsets: each firm changes its compliance behavior until the marginal costs of the various compliance activities are equal. With offsets, firms buy and sell permits, buy and sell offsets, and abate until their marginal abatement cost is equal to the price of a permit and the price of an offset. The equilibrium condition is

$$MAC_i(e_i) = p_p^* = p_o^*.$$

Figure 10.3 illustrates the equilibrium with permits and offsets. Firms use $\bar{e}$ permits and o offsets, and total emissions are $\bar{e} + o$. The policy maker sets $\bar{e}$, and o is determined by setting $S(o) = MAC_M(\bar{e} + o)$.[12] The marginal abatement cost is $MAC_M(\bar{e} + o)$, which is equal to p_2. The equilibrium price of a permit and the equilibrium price of an offset are equal to each other, and equal to the marginal abatement cost, $p_2 = p_o^* = p_p^*$.

The addition of offsets leads to some important differences in the functioning of the tradable permit system, as one can see by comparing figures 10.3 and 10.2. With offsets and permits, total emissions are higher (although any increase in emissions is offset outside of the emissions-control system), the net contribution to pollution is the same, and the market marginal abatement cost is lower. With permits alone, total emissions are $\bar{e}$; with permits and offsets, emissions are $\bar{e} + o$. Likewise, the marginal abatement cost and the price of a permit is p_1 without offsets, but with offsets these are reduced to p_2.

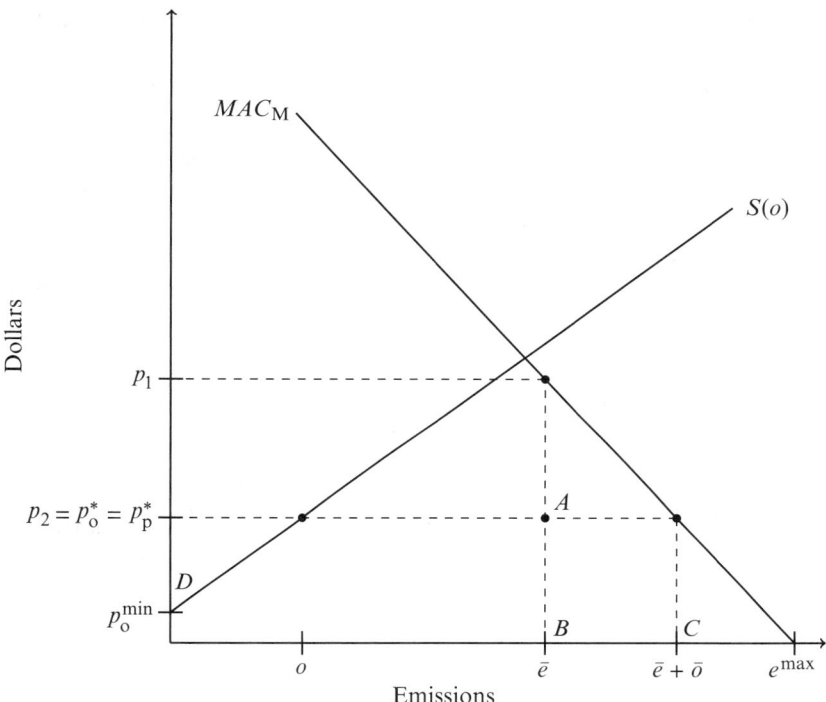

Figure 10.3
Market equilibrium with tradable permits and offsets.

Adding offsets reduces the total compliance costs for regulated firms. Without offsets, total abatement costs are

$$\int_{\bar{e}}^{e^{max}} MAC_M(e)de = Area(A+B+C).$$

With offsets, total abatement costs are

$$\int_{\bar{e}+o}^{e^{max}} MAC_M(e)de = Area(C).$$

After accounting for offset revenues of $p_2 o = Area(B)$, there is still a compliance cost saving of

$$\int_{\bar{e}}^{\bar{e}+o} (MAC_M(e)-p_2)de = Area(A).$$

This analysis is based on the presumption that the policy maker gives the permits away. If instead the policy maker auctions all permits to firms, there is a loss in permit revenue of $(p_1 - p_2)\bar{e}$. The loss in permit revenue is a loss to the policy maker and a decrease in compliance costs for the firms participating in the emissions-control system.

One important issue regarding offsets is that of additionality. For instance, concerns over the non-additionality of offsets issued by reductions in HFC-23 and other industrial gasses prompted the European Commission to ban certain offsets starting in 2013 (Trotignon 2012). So long as offsets are truly additional, the aggregate level of emissions does not change with their introduction even though the level of emissions within the emissions-control system increases by o. In some cases, offsets might reduce emissions by more than their notational reduction if they are additional and their baseline is understated (Bento, Kanbur, and Leard 2012).

10.2 Permit-offset price differential

In the analysis presented in the preceding section, permits and offsets are perfect substitutes and we find that in equilibrium the prices of offsets and permits are equal. To motivate why the equilibrium permit price is greater than the price of offsets, we offer two different but complementary explanations. The first explanation attributes the price differential to quotas placed on offset usage. The second explanation is that there are higher transaction costs associated with offset purchases.

First we ignore transaction costs associated with offsets and just consider a binding offset usage cap or quota. In this case, each firm's marginal abatement cost is equal to the price of a permit; however, because the quota on offsets is binding, the equilibrium price of an offset is not equal to the marginal abatement cost and price of a permit. Figure 10.4 illustrates this case when the offset binding offset quota is $\bar{o}$. The number of offsets surrendered is the maximum number allowed, $\bar{o}$, and total emissions are $\bar{e} + \bar{o}$. The price of an offset is determined by the offset supply function, and permits are traded until the permit price is equal to the market marginal abatement cost. If allowed to do so, firms would use more offsets, and the offset price would increase while the permit price and MAC_M would decrease.

An alternative explanation for a price gap between permits and offsets is the presence of additional transaction costs for offsets. The types of costs we are considering have to do with the transactions to surrender offsets in addition to permits. These additional costs might take place in the form of additional brokerage costs, informational costs, or perhaps reputational costs associated with surrendering offsets. These costs are distinct from costs associated with the creation of offset credits (Michaelowa and Jotzo 2005), which we expect to be

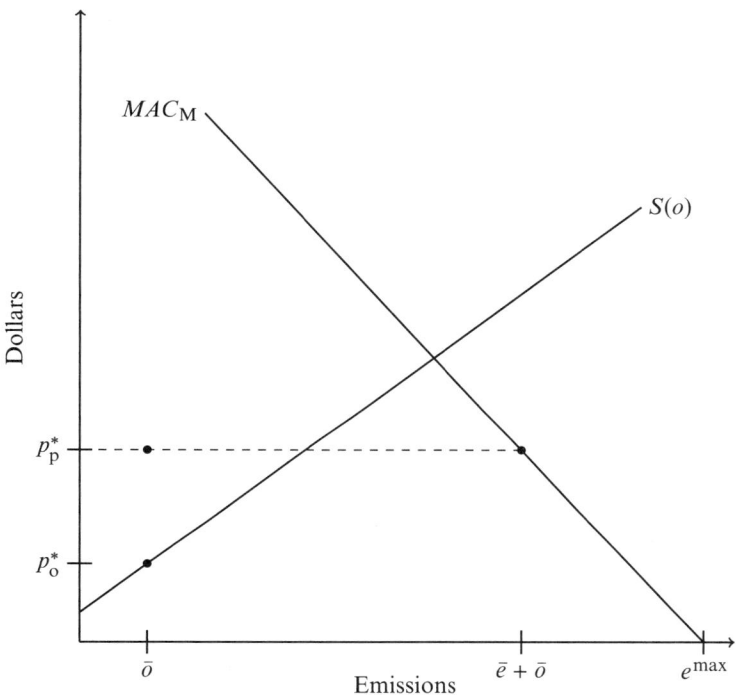

Figure 10.4
Market equilibrium with tradable permits and offsets: Offset usage caps create a price differential.

captured in the exogenous supply curve for offsets. For instance, the additional costs associated with the creation of "Gold Standard" CERs are assumed to be included in the supply curve for CERs and not in the transactional cost associated with surrendering offsets.

We model offset transaction costs in a way similar to that used by Stavins (1995): adding marginal transaction costs associated with offsets to the market supply of offsets. With offset transaction costs, the marginal cost of surrendering an offset is the price of an offset, p_o, plus the marginal transaction cost, $\partial x / \partial o$.

Figure 10.5 depicts the equilibrium with offset transaction costs. A total of o offsets are surrendered, and overall emissions are $\bar{e} + o$. The equilibrium offset price is p_o^* and the overall marginal cost associated with surrendering an offset is $p_o^* + \partial x / \partial o$. In equilibrium, the marginal cost associated with the various abatement activities is equalized, and we have the equilibrium condition

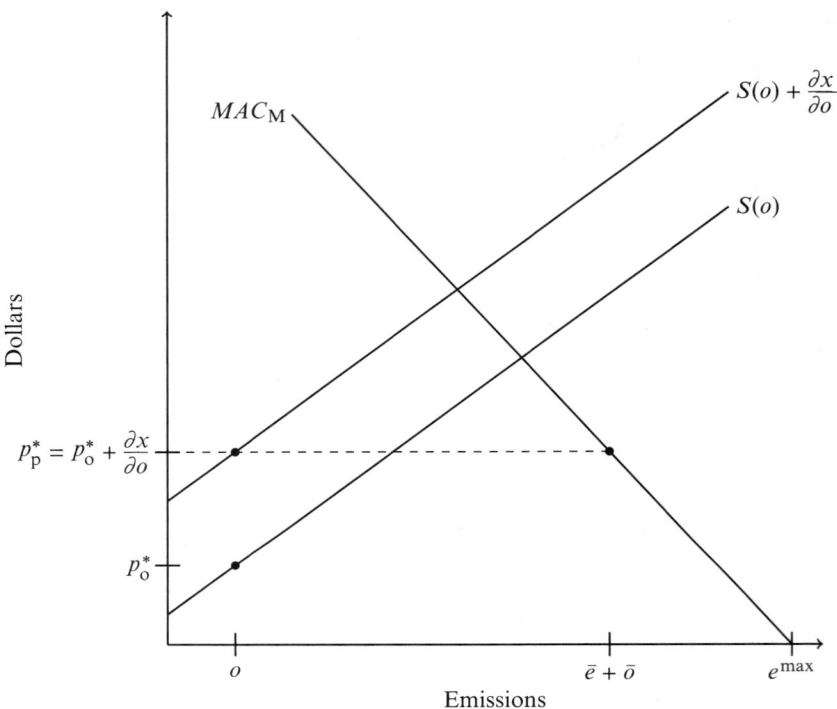

Figure 10.5
Market equilibrium with tradable permits and offsets: Offset transaction costs create a price differential.

$$MAC_M(\bar{e} + o) = p_p^* = p_o^* + \frac{\partial x}{\partial o}.$$

The marginal transaction cost of offsets creates a wedge between the permit price and the price of offsets so that the price of an offset is less than the permit price, $p_o^* < p_p^*$.

The equilibrium depicted in figure 10.5 is analogous to the equilibrium depicted in figure 10.4 with a binding offset cap and leads us to the following result.

Result 1 *A price differential between permits and offsets is caused by either an offset usage quota or additional transaction costs associated with surrendering offsets.*

Comparing either the transaction-cost equilibrium depicted in figure 10.5 or the quota equilibrium in figure 10.4 against the equilibrium without transaction costs or quotas (figure 10.3), we observe that the equilibrium price of offsets is lower and the permit price is higher. Total emissions are lower because firms would buy more offsets in the absence of transaction costs or quotas. Lower emissions imply that firms' abatement costs are greater. Offset revenues are lower because the offset price is lower and fewer offsets are surrendered, also leading to lower offset producer surplus.

10.3 Firm-level offset differences

It was shown in section 10.2 that either offset usage quotas or offset transaction costs cause the price of a permit to be greater than the price of an offset. Each equilibrium depicted with offsets thus far has all firms surrendering the same amount of offsets. In figure 10.3 and in figure 10.5, each firm surrenders o offsets (although o may differ across figures); in figure 10.4, each firm surrenders $\bar{o}$ offsets. Now we explore what might cause different firms to surrender different amounts of offsets.

One possibility is to have firms face different offset usage quotas. With different offset usage quotas, firms surrender different amounts of offsets as long as the price of a permit is greater than the price of an offset. A drawback with this modeling approach is that all firms surrender offsets and each firm surrenders its quota of offsets. In the next section, we show that this result is not consistent with empirical evidence. Instead of allowing for different offset usage quotas, we allow for different transaction costs and then combine this analysis with an offset usage quota.

Firms' heterogeneity in offset usage is modeled by allowing for firms to have different transaction costs associated with offsets. Similar to the discussion of figure 10.2 that allows for firms to have different marginal abatement costs to explain permit trading, we assume there are two types of firms based on their transaction costs associated with offsets. A fraction $\beta \in (0, 1)$ of firms have high marginal transaction costs while $1 - \beta$ firms have low marginal transaction costs. For simplicity we normalize the transaction cost of the $1 - \beta$ low-transaction-cost firms to zero.

The β firms with high transaction costs have $S(o) + \partial x^H / \partial o$ as their marginal cost curve of surrendering offsets. $S(o)$ is the marginal cost curve of surrendering offsets for the $1 - \beta$ firms with no transaction costs. Aggregating the high and low marginal cost curves yields the market marginal cost curve of surrendering offsets, $S(o) + \partial x^M / \partial o$. The

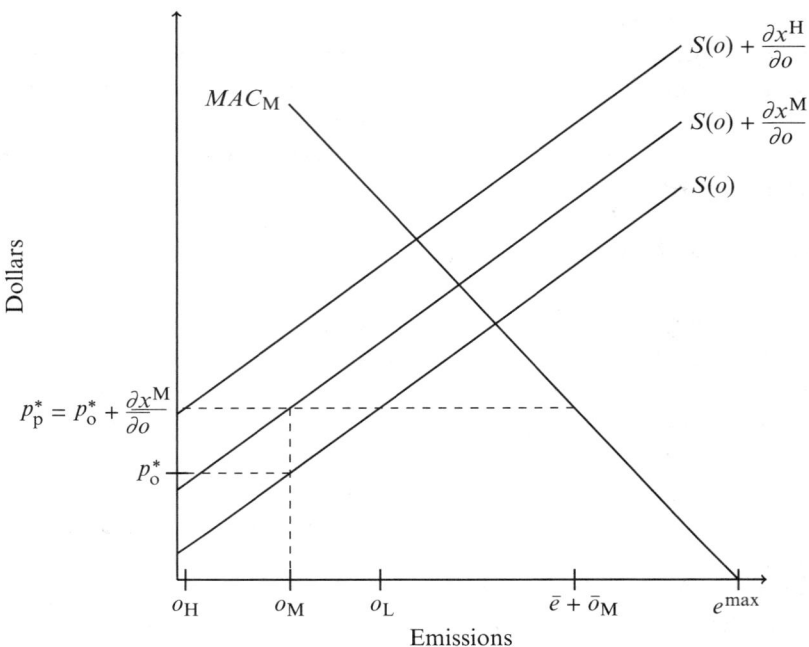

Figure 10.6
Market equilibrium with heterogeneous transaction costs.

resulting equilibrium, illustrated in figure 10.6, corresponds to the equilibrium depicted in figure 10.5.

In equilibrium, the amount of offsets a firm surrenders depends on the transaction costs it faces. Firms with low transaction costs surrender more offsets than firms with high transaction costs. Each of the $1 - \beta$ firms with no transaction costs surrenders o_L offsets, and o_H offsets are surrendered by each of the β high-transaction-cost firms.[13] The total number of offsets surrendered is $\beta o_H + (1 - \beta) o_L = o_M$. The amount o_M in figure 10.6 corresponds to o in figure 10.5. Overall emissions are $\bar{e} + o_M$, and the equilibrium offset price is $p_o^* = S(o_M)$. The marginal cost associated with the various abatement activities is equalized where

$$MAC_M(\bar{e} + o_M) = p_p^* = p_o^* + \frac{\partial x^M}{\partial o_M}.$$

Consider an offset usage quota combined with firm-level differences in offset transaction costs. It is possible to consider firm-level differences in the offset usage quotas, but for simplicity only consider a single quota applicable to all firms. For the quota to be binding for some firms the quota of $\bar{o}$ must be less than o_L. If $\bar{o} \leq o_H$, then the quota is binding for all firms and this situation is depicted in figure 10.4. For an offset quota of $o_H < \bar{o} < o_L$, firms with low transaction costs are at their offset cap while firms with high transaction costs are not. The total number of offsets surrendered now becomes

$$\beta o_H + (1 - \beta)\bar{o} \leq o_M.$$

With firm-level differences in offset transaction costs and an offset usage quota, it is possible to observe three different classifications of firm behavior. One group of firms may have high offset transaction costs such that they never surrender offsets. Another group of firms may have sufficiently low transaction costs relative to their offset quota so that they surrender offsets up to their offset cap. The third group of firms falls in the middle of the first two groups. Their transaction costs are not so high that they surrender some offsets, but their transaction costs are high enough so that they do not surrender up to their offset quota. These three classifications correspond to what has been observed empirically.

The above analysis is summarized by the following result.

Result 2 *Firms surrender different amounts of offsets if there are firm-level differences in offset usage quotas or firm-level differences in transaction costs associated with surrendering offsets.*

With differences only in offset usage quotas, all firms surrender offsets up to their quota. Because the quotas are different, firms surrender different amounts of offsets. This results in just one of the three different classifications of firms described above. With differences in offset transaction costs and an offset usage quota, it is possible to have firms that do not surrender any offsets, firms that surrender offsets but are below their quota, and firms that surrender offsets up to their quota.

10.4 Analysis of the EU ETS

Annual compliance data from the EU ETS are used to relate observed outcomes back to the theoretical implications derived from our model. The data come from the EU Transaction Log (EUTL) and each county's Phase II national allocation plan (NAP).[14] From the EUTL, we obtained annual installation level accounting information for all operator holding accounts. For each installation-year, we observe the allocation of EUAs, verified emissions, the amount of EUAs surrendered, the amount of CERs surrendered, and the amount of ERUs surrendered.[15]

Our other source of data is each country's Phase II NAP. Each NAP provides country-level information regarding the maximum amount of offsets each installation is allowed to surrender to cover verified emissions, and the rules regarding the banking/borrowing of their annual offset constraint. Table 10.7 summarizes this information from each NAP. Most countries specify the maximum amount of offsets allowed as a percentage of the allocation of EUAs. These percentages differ across countries and some countries set different percentages for different sectors. For example, the United Kingdom allows each installation to surrender offsets up to 8 percent of its Phase II allocation of EUAs. but for the combustion sector the relevant percentage is 9.3. The extreme cases are Sweden and Denmark, which set a specific maximum amount of offsets for each installation.

During Phase I and Phase II (that is, during the years 2005–2012), we observe 14,774 regulated installations in 30 of 31 participating countries.[15] From the 14,774 installations we omit 38 installations: all those in Iceland, Liechtenstein, and Malta.[17] We also omit the 1,163 aircraft operators, because their inclusion into the ETS started in 2012.[18] Further restricting the data to include only fully compliant installations with positive verified emissions results in omitting an additional 1,085 installations. After these restrictions, 12,488 installations from 27 countries remain, with a total of 77,049 installation-year observations.

Table 10.2 provides an overview of the average annual behavior in each country during Phase II. Information regarding each country's annual emissions cap in table 10.2 (and table 10.3) comes directly from table 2 in Mansanet-Bataller and Pardo 2008.[19] Both the average annual emissions cap and the average annual reported emissions are listed in million metric tons of CO_2 equivalent. Permits and offsets are reported in millions of units. Surrendered offsets for Estonia are zero because Estonia is the only country that does not allow installations to surrender offsets. All other zeros for surrendered offsets are due to rounding.

In table 10.2, for all countries, the reported emissions are below the emissions cap because installations bank permits into Phase III (2013–2020). To bank permits into the next phase, installations reduce their emissions by more than they otherwise would to create an excess stock of permits to bank and use in the future. The price of a permit at the end of Phase II is the opportunity cost of banking a permit into Phase

Table 10.2
Phase II average annual outcomes. (Reported measures are in million metric tons of CO equivalent per year.)

	Emissions cap	Reported emissions	Surrendered permits	Surrendered offsets
Austria	31	30	27	3
Belgium	59	43	41	2
Bulgaria	42	35	30	4
Cyprus	5	4	3	0
Czech Republic	87	75	67	8
Denmark	25	23	21	2
Estonia	13	10	10	0
Finland	38	32	30	3
France	133	105	91	14
Germany	453	447	388	59
Greece	69	60	56	5
Hungary	27	16	15	1
Ireland	22	18	16	1
Italy	196	180	163	16
Latvia	3	2	2	0
Lithuania	9	2	2	0
Luxembourg	3	2	2	0
Netherlands	86	81	75	6
Norway	n.a.	12	11	1
Poland	209	196	178	19
Portugal	35	26	24	3
Romania	76	49	45	6
Slovakia	33	22	20	2
Slovenia	8	8	7	1
Spain	152	121	105	16
Sweden	23	16	15	1
UK	246	235	220	15
Total	2,081	1,853	1,665	189

sources: EUTL, NAPs, Mansanet-Bataller and Pardo 2008

III.[20] Rather than banking excess permits, one might argue that the emissions cap was set too high (or that realized macroeconomic conditions during Phase II vastly differed from expectations) creating an excess of permits. In this case, the price of a permit should converge to zero as the supply of permits exceeds the demand. Unlike the permit price at the end of Phase I or the offset price at the end of Phase II, the permit price at the end of Phase II did not converge to zero.

Table 10.3
Comparison of Phase I and Phase II. (Reported emissions are in million metric tons of CO_2 equivalent per year. Totals exclude Bulgaria, Cyprus, Norway, and Romania.)

	Phase I emissions	Phase II emissions	Percentage change in emissions	Percentage change in emissions cap
Austria	32	30	−8.2	−7
Belgium	54	43	−21	−5.8
Bulgaria	n.a.	35	n.a.	0
Cyprus	n.a.	4	n.a.	−3.9
Czech Republic	58	75	28.1	−11.1
Denmark	30	23	−22.3	−26.9
Estonia	10	10	2.2	−33.1
Finland	40	32	−20.5	−17.4
France	88	105	19.5	−15.1
Germany	464	447	−3.7	−9.2
Greece	71	60	−15.3	−7.1
Hungary	22	16	−24.1	−14.1
Ireland	21	18	−17.6	0
Italy	174	180	3	−12.2
Latvia	3	2	−28.4	−25.4
Lithuania	6	2	−61.2	−28.5
Luxembourg	3	2	−19.5	−26.5
Netherlands	79	81	2.8	−10
Norway	n.a.	12	n.a.	n.a.
Poland	207	196	−5.4	−12.8
Portugal	34	26	−21.2	−10.5
Romania	n.a.	49	n.a.	1.5
Slovakia	17	22	33.2	6.9
Slovenia	9	8	−7.9	−5.7
Spain	177	121	−31.4	−12.7
Sweden	19	16	−13.7	−0.4
UK	246	235	−4.7	0.4
Total	1,866	1,753	−6	−9.6

sources: EUTL, NAPs, Mansanet-Bataller and Pardo 2008

Also of interest in table 10.2 is that offsets make up a small proportion of surrendered compliance assets. This observation is surprising because the price of a permit is always above the price of an offset, indicating that it is more cost effective to surrender offsets. Table 10.4 shows that most installations do not reach their offset quota and it is possible for these installations to surrender more offsets. With non-

Table 10.4
Offset behavior. (Reported measures are percentages of installations.)

	No offsets	Some offsets	At offset quota
Austria	20	27	53
Belgium	35	42	23
Bulgaria	30	42	27
Cyprus	0	73	27
Czech Republic	28	22	50
Denmark	50	21	29
Finland	54	30	16
France	22	31	47
Germany	23	37	40
Greece	25	22	53
Hungary	49	26	25
Ireland	52	41	7
Italy	24	43	33
Latvia	92	8	0
Lithuania	83	14	3
Luxembourg	23	31	46
Netherlands	28	19	53
Norway	51	38	11
Poland	12	27	61
Portugal	34	29	36
Romania	27	19	54
Slovakia	36	15	49
Slovenia	24	22	53
Spain	20	59	21
Sweden	50	8	41
UK	31	55	14
Total	29	35	36

sources: EUTL and NAPs

binding offset quotas, one possible explanation for why installations do not surrender more offsets is offset transaction costs.

Table 10.3 compares the change in emissions between Phase I and II against the percentage change in the emissions cap across phases.[21] Two observations from this table are that Phase II emissions were lower than Phase I emissions for all but five countries, and average annual total emissions were lower in Phase II. Ignoring for the moment any potential effect of allowing installations to surrender offsets during Phase II, there are many reasons to expect Phase II emissions to decrease relative to emissions in Phase I. Probably the most obvious reason is that emission caps were lower for Phase II. Table 10.3 shows that for all but five countries Phase II emissions caps were lower and the overall average annual emissions cap was lower by 9.6 million metric tons. As has already been mentioned, another potential reason Phase II emissions were lower is because of an installation's ability to bank excess permits into Phase III. It is also possible that the overall economic climate was vastly different during Phase II (2008–2012) than Phase I (2005–2007), resulting in lower emissions in Phase II due to lower levels of economic output. Lower levels of economic output would be represented by a shift inwards of aggregate MAC during Phase II relative to Phase I.

Figure 10.3 shows that when offsets are added to the model, emissions increase from $\bar{e}$ to $\bar{e} + o$. All else equal, we would expect emissions to increase in Phase II when compared with Phase I due to the inclusion of offsets in Phase II and expansion of covered countries. From table 10.3, we find that Phase II emissions were 6 percent less than Phase I emissions. This change in the opposite direction of what the theory from section 10.1 suggests is probably due to changes between Phase I and Phase II.

In table 10.3, the percentage change in average annual emissions across phases is presented next to the percentage change in the emissions cap. The emissions cap includes all emissions regardless of the type of compliance asset surrendered (i.e., the cap is for permits and offsets), and between Phase I and II this cap decreased by 9.6 percent. With other factors ignored for the moment, if firms surrender all the permits and offsets they can surrender during Phase II, we expect their emissions to decrease by 9.6 percent because of the lower overall cap in Phase II. We might expect emissions to decrease by 9.6 percent or more for three reasons: the lower emissions cap, the ability to bank permits into future periods, and reduced economic activity decreasing aggregate demand for emissions.

The fourth column in table 10.3 indicates the opposite. Rather than Phase II emissions decreasing by more than 9.6 percent, overall emissions decreased by only 6 percent. One explanation for the smaller decrease in emissions is that the inclusion of offsets gives firms access to less expensive forms of compliance, which allows them to emit more than in the case with no offsets. Though this is not a precise counterfactual, the empirical evidence is consistent with the theory.

Tables 10.4 and 10.5 show that quotas alone are not sufficient for explaining observed behavior within the EU ETS. Result 2 in section

Table 10.5
Variation in use of offsets. (Means and standard deviations are reported only for installations that surrender offsets but are not at their quota.)

	Some offsets	Mean percentage of quota	Standard deviation percentage of quota
Austria	27	75	28
Belgium	42	44	23
Bulgaria	42	73	31
Cyprus	73	32	4
Czech Republic	22	75	31
Denmark	21	56	28
Finland	30	74	25
France	31	80	25
Germany	37	72	27
Greece	22	69	33
Hungary	26	84	26
Ireland	41	47	18
Italy	43	54	27
Latvia	8	51	56
Lithuania	14	50	61
Luxembourg	31	94	4
Netherlands	19	76	29
Norway	38	65	22
Poland	27	86	24
Portugal	29	74	29
Romania	19	76	26
Slovakia	15	66	38
Slovenia	22	87	19
Spain	59	44	35
Sweden	8	64	36
UK	55	81	16
Total	35	67	31

sources: EUTL and NAPs

10.3 states that either heterogeneity in offset quotas or offset transaction
costs explains why firms surrender different amounts of offsets. Without
offset transaction costs and with different offset quotas, each firm surrenders its quota of offsets (as long as the price of offsets is below the
price of permits). Because the quotas differ across firms, different firms
surrender different amounts of offsets.

Whereas table 10.7 shows that there is heterogeneity in offset quotas,
table 10.4 indicates that only 36 percent of installations surrender offsets

up to their quota. The evidence presented in table 10.4 is not consistent with just offset quotas, but is consistent with firms' having offset quotas and offset transaction costs. This leads us to the following result.

Result 3 *Allowing only for differences in offset usage quotas is not consistent with the empirical evidence, whereas allowing for differences in offset transaction costs along with offset usage quotas is consistent with empirical evidence.*

According to table 10.4, 29 percent of installations do not surrender any offsets. In the framework of our model, we consider these firms to have prohibitively high offset transaction costs. In contrast, 36 percent of installations have sufficiently low offset transaction costs, because these installations surrender their full quota of offsets.

The remaining 35 percent of installations surrendering some offsets, but not their full quota, are further outlined in table 10.5. Rather than surrendering 100 percent of their offset quota, these installations surrender, on average, only 67 percent of their offset quota. Note that this average varies across countries and that there is significant variation within each country.

Conclusion

This chapter extends the textbook model of tradable emissions permits to allow for two kinds of compliance assets (permits and offsets). The static model developed in section 10.1 allows for the supply of permits to be endogenously determined by the heterogeneity of abatement costs relative to permit endowment, but the supply of offsets is assumed to be exogenous. Two results of the model are that in equilibrium the price of permits and offsets are the same, and that all firms surrender the same amount of offsets.

We extend the model to show how it can be consistent with observed outcomes in the EU ETS. Section 10.2 offers two explanations for a persistent price difference between two types of compliance assets that are perfect substitutes when they are surrendered. Result 1 indicates that a price gap between permits and offset could be caused by either offset usage quotas or offset transaction costs.

Section 10.3 further embellishes the model by introducing the possibility of heterogeneous transaction costs. From Result 2, firms surrender different amounts of offsets with either different offset quotas or heterogeneity in offset transaction costs. Result 3 further indicates that offset usage quotas alone cannot explain why firms surrender different amounts of offsets. Our extensions of the model provide theoretical underpinnings for observed outcomes in the EU ETS.

Supplemental materials

Table 10.6 describes how the number of installations changed between Phase I and Phase II. Table 10.7 describes the offset restrictions that were in place for Phase II of the EU ETS.

Table 10.6
Average annual number of installations for each phase. (Totals exclude Bulgaria, Cyprus, Norway, and Romania.)

	Number of Phase I installations	Number of Phase II installations
Austria	192	197
Belgium	304	277
Bulgaria	n.a.	123
Cyprus	n.a.	8
Czech Republic	270	355
Denmark	355	334
Estonia	42	35
Finland	457	468
France	777	905
Germany	1,726	1,600
Greece	141	125
Hungary	212	167
Ireland	107	100
Italy	683	983
Latvia	87	66
Lithuania	91	63
Luxembourg	15	13
Netherlands	204	368
Norway	n.a.	86
Poland	820	765
Portugal	230	171
Romania	n.a.	210
Slovakia	119	160
Slovenia	94	88
Spain	867	928
Sweden	566	463
UK	697	872
Total	9,057	9,504

source: EUTL

Table 10.7
Offset rules from NAPs. (Allocation percentage refers to the amount of offsets firms can surrender as the percentage of their EUA allocations.)

	Allows offset banking	Allows offset borrowing	Allocation percentage	Notes and exceptions
Austria	x	x	10	
Belgium				
Brussels	x	x	50	
Flanders	x	x	7	24 for Combustion Sector
Wallonia	x	x	4	
Bulgaria	x	x	12.51	
Czech Republic	x	x	10	
Denmark	x	x	NA	Installation Specific
Finland	x	x	10	
France	x	x	13.5	
Germany	x	x	22	
Greece	x	x	9	
Hungary			10	
Ireland	x	x	5	11 for Combustion and Cement/Lime Sector
Italy	x		7.5	19.3 for Combustion Sector, 16.7 for Steel/ Iron Sector
Latvia			10	
Lithuania			10	
Luxembourg	x	x	10	
Netherlands	x	x	10	
Norway	x		20	
Poland	x		10	
Portugal	x	x	10	
Romania	x	x	10	
Slovakia	x	x	7	
Slovenia	x	x	15.76	
Spain	x		7.9	42 for Combustion Sector
Sweden	x	x	NA	Installation Specific
UK	x		8	9.3 for Combustion Sector

Acknowledgments

We thank Marc Gronwald, Beat Hintermann, Frank Jotzo, and seminar participants at the CESifo Venice Summer Institute 2013 for helpful comments. We also thank James Banovetz for excellent research assistance.

Notes

1. There is also a large variance in the amount of offsets an installation is allowed to surrender. In table 10.5, we show that there is still a large amount of variance in the amount of offsets surrendered controlling for variation in offset quotas.

2. With regards to the EU ETS, we refer to EUAs as permits and both CERs and ERUs as offsets throughout the chapter. Figure 10.1 uses EUA and secondary CER (sCER) prices as CERs are more prevalent than ERUs. It is important to note that the price of "Gold Standard" CERs has been higher than the price of EUAs, but these offsets involve additional certification costs.

3. The source of these data is discussed in further detail in section 10.4.

4. In this chapter firms and installations are interchangeable, although a firm may be a collection of installations.

5. Both transaction costs and institutional rules (offset quotas being an institutional rule) are mentioned by Convery (2009) as areas worth considering for future research.

6. Chevallier 2011, Convery and Redmond 2007, and Ellerman and Buchner 2008 serve as useful references.

7. See Chevallier 2010; Conrad, Rittler, and Rotfu 2012; Gronwald, Ketterer, and Trück 2010; Mansanet-Bataller, Chevallier, Hervé-Mignucci, and Alberola 2011; Mizrach 2012; Nazifi 2013.

8. Löschel, Gallier, Lutz, Brockmann, and Kieckhöner (2013) include recent survey evidence indicating abatement cost heterogeneity in the EU ETS. In particular, smaller firms face much higher abatement costs.

9. Differences in permit allocation mechanisms are discussed in section 10.4.

10. Trade could also occur if there is heterogeneity in the allocation of permits.

11. In practice, there may be differences between permits and offsets due to institutional rules and firms' perceptions. We consider these differences in section 10.2 by adding offset quotas and transaction costs associated with offsets to the model.

12. It is possible that the price intercept of the offset supply curve, p_o^{min}, is greater than p_1. In this case, no offsets are surrendered and introducing offsets does not change the equilibrium. Also note that the intersection of offset supply and the aggregate marginal abatement cost is not informative for the problem.

13. Note that if transaction costs are prohibitively high some firms will not surrender any offsets.

14. The EUTL can be found at http://ec.europa.eu/environment/ets/. Each country's NAP can be found at http://ec.europa.eu/clima/policies/ets/pre2013/nap/documentation_en.htm.

15. CERs and ERUs are combined in our measure of the amount of offsets surrendered.

16. No installation data are available for Gibraltar, a dependent territory of the UK.

17. Iceland was not fully participating in the ETS, and we do not have sufficient offset information from the NAP for the remaining eight installations in Liechtenstein and Malta. Note that Malta and Cyprus are EU members, and that Liechtenstein, Iceland, and Norway cooperate in the EU ETS but are not members of the EU.

18. Aircraft operators are also allowed to surrender European Union Aviation Allowances (EUAAs) to cover verified emissions. Because EUAAs have only been valid since 2012, we exclude them from our analysis.

19. The emissions cap for Norway is missing.

20. It is important to note that permits from Phase I and a majority of offsets from Phase II could not be banked into the subsequent phase. Because these compliance assets held no future value, their prices converged toward zero at the ends of their respective phases.

21. Phase I emissions for Bulgaria, Cyprus, Norway, and Romania are not included, as these countries did not participate fully during Phase I of the EU ETS.

References

Alberola, E., J. Chevallier, and B. Cheze. 2008. Price drivers and structural breaks in European carbon prices 2005–2007. *Energy Policy* 36 (2): 787–797.

Anger, N. 2008. Emissions trading beyond Europe: Linking schemes in a post-Kyoto world. *Energy Economics* 30 (4): 2028–2049.

Bento, A., R. Kanbur, and B. Leard. 2012. Super-additionality: A neglected force in markets for carbon offsets. Discussion paper 8952, Centre for Economic Policy Research.

Chevallier, J. 2010. EUAs and CERs: Vector autoregression, impulse response function and cointegration analysis. *Economic Bulletin* 30 (1): 558–576.

Chevallier, J. 2011. Carbon price drivers: An updated literature review. *International Journal of Applied Logistics* 4 (4): 1–7.

Conrad, C., D. Rittler, and W. Rotfu. 2012. Modeling and explaining the dynamics of European union allowance prices at high frequency. *Energy Economics* 34 (1): 316–326.

Conte, M. N., and M. J. Kotchen. 2010. Explaining the price of voluntary carbon offsets. *Climate Change Economics* 1 (02): 93–111.

Convery, F. J. 2009. Reflections—the emerging literature on emissions trading in Europe. *Review of Environmental Economics and Policy* 3 (1): 121–137.

Convery, F. J., and L. Redmond. 2007. Market and price developments in the European Union Emissions Trading Scheme. *Review of Environmental Economics and Policy* 1 (1): 88–111.

Creti, A., P.-A. Jouvet, and V. Mignon. 2012. Carbon price drivers: Phase I versus Phase II equilibrium? *Energy Economics* 34 (1): 327–334.

Ellerman, A., and B. Buchner. 2008. Over-allocation or abatement? A preliminary analysis of the EU ETS based on the 2005–06 emissions data. *Environmental and Resource Economics* 41 (2): 267–287.

Fankhauser, S., and C. Hepburn. 2010. Designing carbon markets. Part I: Carbon markets in time. *Energy Policy* 38 (8): 4363–4370.

Flachsland, C., R. Marschinski, and O. Edenhofer. 2009. To link or not to link: Benefits and disadvantages of linking cap-and-trade systems. *Climate Policy* 9 (4): 358–372.

Gronwald, M., J. Ketterer, and S. Trück. 2010. On the origins of emission allowance price fluctuations. Presented at International Energy Workshop, Stockholm.

Heindl, P. 2012a. Financial intermediaries and emissions trading: Market development and pricing strategies. Discussion paper 12-064, Zentrum für Europäische Wirtschaftsforschung.

Heindl, P. 2012b. Transaction costs and tradable permits: Empirical evidence from the EU Emissions Trading Scheme. Discussion paper 12-021, Zentrum für Europäische Wirtschaftsforschung.

Hintermann, B. 2010. Allowance price drivers in the first phase of the EU ETS. *Journal of Environmental Economics and Management* 59 (1): 43–56.

Jaraitė, J., F. Convery, and C. Di Maria. 2010. Transaction costs for firms in the EU ETS: Lessons from Ireland. *Climate Policy* 10 (2): 190–215.

Jaraitė-Kažukauske, J., and A. Kažukauskas. Forthcoming. Do transaction costs influence firm trading behavior in the European emissions trading system? *Environmental and Resource Economics*. doi:10.1007/s10640-014-9831-7.

Kettner, C., D. Kletzan-Slamanig, and A. Köppl. 2012. The EU Emission Trading Scheme: National allocation patterns and trading flows. Working paper, Austrian Institute of Economic Research.

Klepper, G. 2011. The future of the European emission trading system and the clean development mechanism in a post-Kyoto world. *Energy Economics* 33 (4): 687–698.

Klepper, G., and S. Peterson. 2006. Emissions trading, CDM, JI, and more: The climate strategy of the EU. *Energy Journal (Cambridge, Mass.)* 27 (2): 1–26.

Löschel, A., C. Gallier, B. Lutz, K. L. Brockmann, and C. Kieckhöner. 2013. KfW/ZEW CO_2 Barometer 2013—Carbon Edition: The EU Emissions Trading Scheme: Firm Behavior During the Crisis (downloaded from https://www.kfw.de).

Lückge, H., and S. Peterson. 2004. The role of CDM and JI for fulfilling the European Kyoto commitments. Working paper 1232, Kiel Institute for the World Economy.

Mansanet-Bataller, M., J. Chevallier, M. Hervé-Mignucci, and E. Alberola. 2011. EUA and sCER Phase II price drivers: Unveiling the reasons for the existence of the EUA-CER spread. *Energy Policy* 39 (3): 1056–1069.

Mansanet-Bataller, M., and Á. Pardo. 2008. What you should know about carbon markets. *Energies* 1 (3): 120–153.

Metcalf, G. E., and D. Weisbach. 2012. Linking policies when tastes differ: Global climate policy in a heterogeneous world. *Review of Environmental Economics and Policy* 6 (1): 110–129.

Michaelowa, A., and F. Jotzo. 2005. Transaction costs, institutional rigidities and the size of the clean development mechanism. *Energy Policy* 33 (4): 511–523.

Mizrach, B. 2012. Integration of the global carbon markets. *Energy Economics* 34 (1): 335–349.

Nazifi, F. 2013. Modelling the price spread between EUA and CER carbon prices. *Energy Policy* 56:434–445.

Stavins, R. N. 1995. Transaction costs and tradeable permits. *Journal of Environmental Economics and Management* 29 (2): 133–148.

Sterk, W., and H. Wang-Helmreich. 2008. Use of external units in the European Union Emissions Trading System post-2012. JIKO policy paper 3/2008.

Trotignon, R. 2012. Combining cap-and-trade with offsets: Lessons from the EU-ETS. *Climate Policy* 12 (3): 273–287.

Tuerk, A., M. Mehling, C. Flachsland, and W. Sterk. 2009. Linking carbon markets: concepts, case studies and pathways. *Climate Policy* 9 (4): 341–357.

Wara, M. 2007. Is the global carbon market working? *Nature* 445 (7128): 595–596.

Wood, P. J., P. Heindl, F. Jotzo, and A. Löschel. 2013. Linking price and quantity pollution controls under uncertainty. Discussion paper 13-025, Zentrum für Europäische Wirtschaftsforschung.

11 Offset Spread Options in the European Carbon Market

Timothy Fitzgerald

In the years 2008–2012, observers of the European Union Emissions Trading System puzzled over the use or non-use of carbon offset credits.[1] Under the rules of the carbon permit scheme, emitters can surrender qualifying offsets in preference to either reducing emissions or surrendering additional tradable allowances. This chapter considers how the nature of the alternative compliance assets accounts for the observed behavior. Emitters prefer to wait to use offsets because the rules governing their use create a spread option that may be valuable to hold as part of a compliance portfolio. Fluctuating asset prices change the value of the option over time, and emitters might wish to time their usage to correspond with the highest value. Faced with the prospect of an expiring option that is in the money, the use of offsets increased later in the Kyoto compliance period, particularly in 2012. The nature of country-specific offset usage rules, which date back to the 1997 Kyoto Agreement, allowed emitters to effectively exchange one asset (tradable allowances) for another (carbon offset credits).[2] As years passed, the number of remaining opportunities to exercise dwindled, and the offset use decision approached an expiration date. As this deadline approached, more offsets were surrendered.

Implementation of offset usage in carbon trading schemes is an important design feature for carbon trading markets. Other tradable permit markets may also encounter this issue. The EU ETS has been a model for carbon trading schemes in Australia, in California, and elsewhere. The EU ETS is currently linked to the emissions trading schemes in Iceland, Liechtenstein, and Norway, with proposed linkages to Australia and Switzerland.[3] The integration of alternative compliance assets is a critical issue for all emissions trading schemes. Optimal design for incorporating offsets has been explored by a number of authors (Bento et al. 2012; Sovacool 2011; Conte and Kotchen 2010). In addition, because of the prominence of the EU ETS for global carbon policy, the ramifications of offsets for the functioning of the EU ETS are also important (Trotignon 2012; Mansanet-Bataller et al. 2011; Trotignon and Leguet 2009).

The approach in this chapter is to consider compliance decisions at the level of the emitting installation. Presented with a choice of compliance strategies, emitters enjoy the option of when to exchange compliance assets, if at all. Although the arbitrage opportunity may be in the money, the option value can make waiting to surrender a more valuable strategy. Particularly as Europe has suffered through nearly continuous economic turmoil during the years of the recent phase (2008–2012), the uncertainty about optimal compliance strategies has been central. Although it is intuitively appealing to estimate the arbitrage gains from exchanging an emissions allowance for an alternative asset, the data requirements for such a calculation are prohibitive. So instead this chapter focuses on critical decision nodes in the compliance process, when employed strategies can be observed.

Spread options arise in a variety of contexts and have attracted considerable interest in the finance literature. The most theoretically appealing way to consider the value of the option is in a continuous setting; the problem is not trivial because the assets may evolve together, requiring the estimation of a non-stable joint distribution. Li et al. (2008) and Venkatramanan and Alexander (2011) have presented closed-form methods for approximating the value of spread options. An alternative continuous approach relies on copulas to construct the joint distribution; Benth and Kettler (2011) used the approach in the context of the European spark spread, and Herath et al. (2013) used a similar model to address crack spread options for oil. Unfortunately all of these models rely on continuous data that are lacking in the EU ETS framework.[4]

The observable data points are years in the compliance period for each compliant installation. There are five nodes: at the end of each year of the compliance period. Although abatement decisions are effectively continuous through the year, the decision about how to comply, or what collection of assets to surrender, is conditional on those decisions because additional curtailment is not possible once emissions are verified. Borovkova et al. (2012) construct a estimable binomial tree for spread and basket options based on a generalized log-normal distribution. The results of this model compare favorably to other estimable models. In order to use the available data to address the value of a spread option between alternative compliance assets and tradable allowances, this chapter employs a trinomial pricing tree. The trinomial tree is known to be a special case of the binomial tree. Because of the relatively small number of nodes, the computational burden imposed by the trinomial is minimal.

Applying an option model to the data on compliance decisions yields a number of interesting results. First, the model explains the temporal pattern of observed compliance behavior. Second, the option model allows for calculation of implied volatilities of the underlying spread. This offers a perspective on market participants' views of the possible exchange. These implied volatilities are noticeably large, reflecting the uncertainty surrounding the surrender of alternative compliance assets throughout the period. Third, recent regulatory changes recognize the peculiar incentives faced by installations with respect to exercising the spread option by surrendering alternative compliance assets. It further suggests that the extended time period for using these assets will lead to a similar dynamic behavior in the future.

11.1 Background

A variety of compliance assets is available, both in primary and secondary markets. Table 11.1 lists the main types of alternative assets along with respective volumes in 2010 and 2011, and how many countries accepted the asset for the registry. The acceptability of these alternative assets varies by country. In writing national allocation plans, some countries adopted more stringent requirements than others.[5] Estonia is

Table 11.1
Non-allowance compliance assets.

Asset type	Volume (million tons of CO2 equivalent)		Number of countries
	2010	2011	
Primary CDM	265	291	20
Secondary CDM	1,275	1,822	20
Primary JI	41	28	20

Note: There are a number of subclassifications of CDM credits, including temporary and long-term credits, not addressed explicitly here. There are also different degrees of certification, such as green and gold standards, which market participants certainly recognize. Emission reduction units (ERUs) are derived from JI projects.
source: Peters-Stanley and Hamilton 2012

the only country that does not accept any offset assets for compliance—all other countries made some provision for accepting offsets, typically ranging between 10 and 20 percent of allocated allowances. Braun et al. (2014) provides additional information about surrendered offsets in each country covered by the ETS.

In July 2013, partly because of the unexpectedly low usage of alternative assets to the registry during the second compliance period, the European Commission issued a draft regulation that relaxed the constraint facing installations. Alternative asset allowances that were unused at the end of the second compliance period were banked forward and will be accepted during the third compliance period (2013–2020), provided that the unused portion did not exceed 11 percent of allocated European Union Allowances (EUAs). Additionally, installations that were not initially permitted to surrender alternative assets were given a quota equal to 4.5 percent of allocated EUAs. Commercial airlines, which did not come under the purview of the EU ETS until 2012, were also granted a quota of 1.5 percent of EUA allocation. These quotas can be used during the years 2013–2020. This draft regulation is still under debate in the EU Parliament, but it represents a drastic ex-post relaxation of the constraint on surrender of alternative assets.

11.2 Model

Profit-maximizing emitters must minimize costs of compliance with the emissions system. In each time period, the unregulated emitter chooses an output level (y) and a level of abatement (a). The emissions technology, which may be specific to the individual emitter, then determines a level of emissions (e) as a function of these two choices in each time period:

$$e_{it} = E_i(y_{it}, a_{it}). \tag{1}$$

For some installations, upper or lower constraints may bind in the choice of output level. Installation capacity puts an upper bound on output, but some installations may be forced to meet minimum output levels. As an example, electric utilities are obligated to provide power. Electric demand can vary substantially with random factors such as weather. To the extent that y_{it} is a function of a random variable that represents weather or other stochastic demand factors, so too are emissions a function of the random variable.[6] Decisions about output and abatement may be effectively continuous, but compliance decisions are discrete. Compliance is enforced annually rather than continuously. This implies that each year emitters determine what portfolio of assets to surrender to cover their emissions. Every emitter faces the same vector of exogenous factor prices $\mathbf{w}$ and faces an exogenous output price q_t.[7] The emitter seeks to choose output and abatement to maximize profits in each period,

$$\max_{y,a} q_t y_{it} - C_i \left(\mathbf{w}_t, y_{it}, a_{it} \right), \tag{2}$$

perhaps subject to capacity constraints that are ignored here. For each firm and time period let $e^* = E(\bar{y}, 0)$ be the unrestricted level of emissions.[8] If each installation emitted at its unregulated level, the overall emissions would be "business as usual."

Once an emissions scheme has been put in place, the emitter still attempts to minimize costs. For simplicity, assume that the probability of detection and penalty for noncompliance are sufficiently high that the emitter always chooses to comply. Assume further that all emitters are small enough to be price takers in the allowance market. The new problem is as follows:

$$\max_{y,a} q_t y_{it} - C_i \left(\mathbf{w}_t, y_{it}, a_{it} \right) - p_t^e E_i \left(y_{it}, a_{it} \right). \tag{3}$$

Under this circumstance familiar first-order conditions arise. The output level is chosen to equate price and the marginal cost of output, now including the emission cost. The optimal abatement decision sets the marginal cost of abatement equal to the price of an additional emissions allowance, holding output constant. Assumptions about the separability of costs are needed to specify the timing of the decisions.[9]

The stochasticity of emissions is likely to be related with overall economic activity.[10] Assuming that abatement technology displays more stickiness than economic activity, it is reasonable to expect p_t^e to be relatively high during periods of high aggregate demand, and lower when economic activity tapers off.

Introducing an alternative compliance asset such as a CDM instrument complicates equation 3 by introducing a decision about how to account for emissions. This is equivalent to introducing an additional decision variable $\alpha \in [0,1]$, which is the proportion of emissions covered using the offset asset. This adds an additional term to the objective function in equation 3, and the emitter now faces

$$\max_{y,a,\alpha} q_t y_{it} - C_i \left(\mathbf{w}_t, y_{it}, a_{it} \right) - p_t^e (1-\alpha) E_i \left(y_{it}, a_{it} \right) - p_t^o \alpha E_i \left(y_{it}, a_{it} \right). \tag{4}$$

This problem yields standard first-order conditions that depend on the prices of each asset. Margrabe (1978) describes the price of an option to exchange one asset for another when the two prices are described by potentially correlated Wiener processes. The choice of which asset to use depends on the prices p^e and p^o. The emitter should favor the lower-priced asset. In the event that the two prices are exactly equal, the emitter is indifferent about which asset to use.

A critical institutional detail is that countries bound by the Kyoto Agreement are obligated to meet targets for emissions reductions, so unlimited use of CDM instruments might create a violation. As a result, all countries have placed limits on the amount or fraction of emissions that a given installation can be covered with assets other than allowances. A handful of small countries (Latvia, Lithuania, and Hungary) opted for an annual limit on emissions, but most others allowed for a phase-long limit. This effectively gives emitters a choice of which of five years in the phase to surrender assets other than allowances.[11] Representing each emitter's individual limit as $\bar{o}_i$, the phase-long limit takes the form

$$\sum_{t=1}^{5} \alpha_{it} E_i(y_{it}, a_{it}) \leq \bar{o}_i. \tag{5}$$

Alternative assets present two choices for the emitter. The first is how many of each type to surrender, or what share of the total allowable amount to use. For emitters that are small relative to the allowance and offset markets, the decision has a bang-bang type solution. If exercise is optimal, then the optimal quantity is to exchange as many allowances as possible. This characteristic leads naturally to the second choice, which is when to exercise the option and surrender offsets. This study is largely concerned with the timing of the spread option decision.

Surrendering an alternative asset into the registry is therefore like exercising an option, with a specified limit on how many option contracts may be surrendered by each installation. Once a credit is surrendered, the emitter cannot take advantage of the option in the future. The option has been exercised. The optimal compliance portfolio decision will therefore take into account the optimal use of all available assets, when the spread option can be exercised at any time up to expiration. This option can be modeled as either an American option or a compound European option.

11.3 Modeling alternatives as options

Margrabe (1978) develops the value of the spread option as a function of the prices of the two assets and time to expiration, designated $\tilde{t}$:

$$V = p^e f \left(\frac{ln\left(\frac{p^e}{p^o}\right) + \frac{1}{2}v^2(\tilde{t})}{v\sqrt{\tilde{t}}} \right) + p^o f \left(\frac{ln\left(\frac{p^e}{p^o}\right) - \frac{1}{2}v^2(\tilde{t})}{v\sqrt{\tilde{t}}} \right), \tag{6}$$

where F is a standard normal distribution. In the event that the price of the alternative asset is higher than the price of an emissions allowance, no installation should choose to exercise the option to surrender an offset instead of an allowance. Borovkova et al. (2012) develop a binomial tree to approximate spread and basket options with potentially correlated log-normal distributions. This makes the estimation of the value of the option estimable under lesser information requirements. This chapter adjusts that methodology slightly to use a trinomial tree.

   Because of the finite number of compliance periods, a trinomial option pricing model is an appropriate model of the value of surrendering alternative compliance assets as opposed to allowances. Observable compliance decisions are made at five nodes. Allowing for a trinomial model of asset pricing follows from the well-known contribution of Cox et al. (1979) with the binomial pricing model. For positive values of $p_t^e - p_t^o = S_t$ the option is in the money; for negative values the option is out of the money. The binomial model allows the price difference to increase or decrease; the trinomial model also allows for the model to remain constant. If the price gap is constant in time, then the decision about when to exercise the option will be governed by the discount rate.

   The optimal timing of exercising the option depends on the price path of the underlying assets. For simplicity, this chapter focuses on the price differential. In principle, the price difference could be either positive or negative, which differs from standard option models in which the price of the underlying asset is non-negative. If offsets were more expensive than allowances (a negative price differential), then they would be unlikely to be an attractive compliance alternative. Assume that the pertinent distribution of price differences is log-normal for analytical simplicity. Suppose that the price difference can either increase, decrease, or remain constant in any given period.[12] This assumption gives rise to a trinomial price tree for the arbitrage. Following Kamrad and Ritchken (1991), consider the logarithm of price one increment of time in the future. The future price can be expressed as a function of the current price

$$\ln(S_{t+1}) = \ln(S_t) + \phi_{t,},$$
(7)

where ϕ_t is a discrete probability density given by

$$\varphi_t = \begin{cases} v & \text{with probability } p_1 = \left(\dfrac{e^{r/2} - e^{-\sigma\sqrt{1/2}}}{e^{\sigma\sqrt{1/2}} - e^{-\sigma\sqrt{1/2}}} \right)^2 \\ 0 & \text{with probability } p_2 = 1 - p_1 - p_3 = 1 - 1/\lambda^2 \\ -v & \text{with probability } p_3 = \left(\dfrac{e^{\sigma\sqrt{1/2}} - e^{r/2}}{e^{\sigma\sqrt{1/2}} - e^{-\sigma\sqrt{1/2}}} \right)^2 \end{cases}$$

with the step $v = \lambda\sigma$ and $\lambda \geq 1$. One advantage to using the trinomial model is that the probabilities of each path can be expressed in terms of observable parameters (Boyle 1986). In the case of compliance assets, there are no dividend yields to complicate valuation. For an exercise step of annual compliance periods, the transition probabilities are strictly a function of the risk-free interest rate, the volatility σ, and the λ scaling parameter. Note that $\lambda = 1$ collapses the trinomial tree to a binomial tree.

11.4 Data

In order to investigate the empirical timing of credit usage, installation-level data from country registries were compiled for years 2008–2012. These data allow analysis of installation-year exercise of the spread option.[13] Table 11.2 provides an overview of use during the Kyoto compliance period. It shows the number of installations that remain open and compliant throughout the period, and that are subject to the phase-long cap, which means that they face the option structure described above.

Participants in the carbon markets have two major options for trading compliance assets. There are market exchanges for carbon assets; over-the-counter (OTC) contracts are also widely used. Both types of trades can be made on a spot or future basis. As future markets become short, we expect convergence of futures and spot prices. In this

Table 11.2
Summary statistics for alternative asset usage. (Cumulative percent allowed surrendered excludes observations that appear to have exceeded limits but are recorded as in compliance.)

	Number of installations	Annual use offsets	Percent users	Percent allowed surrendered
2008	8135	492	6.05	17.4
2009	8135	514	6.32	14.8
2010	8135	1228	15.1	24.5
2011	8135	1549	19.0	33.5
2012	8135	4862	59.8	60.5
Total	8135	5726	70.3	78.1

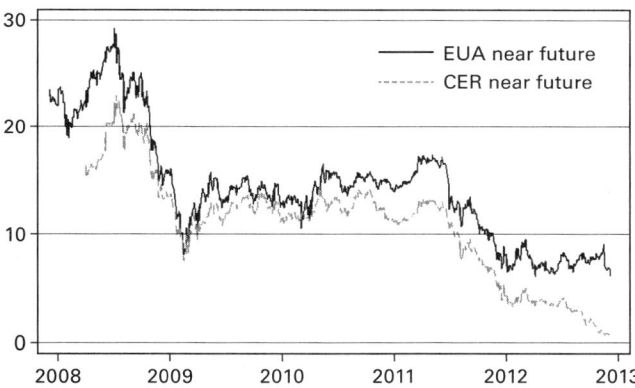

Figure 11.1
EUA and CER near futures: 2008–2012 contracts.

study I use futures contracts from market exchanges. Market observers have noted a shift toward futures contracts, though OTC future contracts are important (Peters-Stanley and Hamilton 2012). Unfortunately OTC prices are not usually observable. Future contracts for emissions allowances and certified emissions reductions are delivered once a year, in early December. Pricing data are from near futures contracts as recorded on the European Energy Exchange.[14] Figure 11.1 depicts the movement in these prices over the years 2008–2012. Figure 11.2 plots the price gap between the two assets over the same years. Trading and price discovery for ERUs has been rather thin over much of the time period; observed prices for ERUs and CERs have been quite close, so the CER price is used as a representation for the alternative asset.

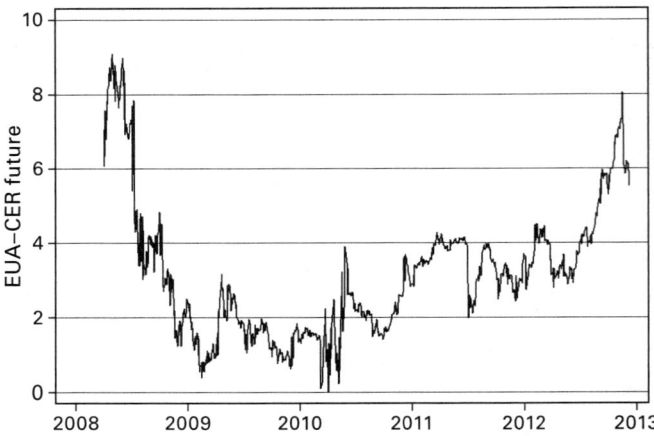

Figure 11.2
Near future spread: 2008–2012 contracts.

Compliance alternatives give installations opportunities for gains beyond the CER futures prices. As an example of one alternative asset class, it is possible that installations can surrender primary CERs. The incentive for doing so would be the verification and delivery risk premium that exists between primary and secondary CERs. Using the relatively safe and deep secondary market effectively constructs a lower bound on the possible gains from exchanging an allowance for an alternative asset. Larger gains from assets such as primary CERs depend in part on the risk premium, which we do not attempt to address.

One shortcoming of the currently available data is that the date of purchase and surrender cannot be observed throughout the phase. This makes it impossible to identify realized gains from surrendering alternative assets instead of allowances. However, by exploiting the reality that the surrender decision is made only once a year, it is possible to estimate the marginal returns under the assumption that the market is at least weakly efficient. The marginal asset surrendered in lieu of an allowance could be purchased at an observable price. Assuming that there are no transaction costs, the value of the arbitrage is simply the spread on the last available day.[15] The exercise values at each of the nodes in the trinomial tree are listed in table 11.3. These values represent the gains from exercising the option by surrendering one alternative asset instead of one allowance.

Table 11.3
Exercise values.

Node	Date	EUA	CER	Spread
Initial spread	3/26/2008	22.47	16.40	6.07
Delivery 2008	11/27/2008	16.27	14.35	1.92
Delivery 2009	11/30/2009	13.11	12.20	0.91
Delivery 2010	11/30/2010	14.77	11.80	2.97
Delivery 2011	11/30/2011	8.35	5.62	2.73
Delivery 2012	11/30/2012	6.22	0.69	5.53

Table 11.4
Historic volatilities: annual.

	EUA	CER	Spread
2008	0.13	0.12	0.49
2009	0.13	0.13	0.39
2010	0.08	0.07	0.56
2011	0.19	0.23	0.17
2012	0.09	0.52	0.27

A second important aspect of the problem is volatility. Table 11.4 shows the historic volatility in a one-year rolling window before each node, which could be used to predict future price movements or evolution of the spread option. At least two important conclusions should be drawn from the table. First, the volatilities change substantially over the course of the phase. Historic volatilities are an important tool for valuing assets, but expected volatility may differ from the historic pattern. A number of individuals could reach different expectations about volatility after examining the same historic record. In the final year we do observe an increase in volatility for the expiring assets, confirming the Samuelson hypothesis.

Second, the volatility of the price spread is far greater than the volatility of the underlying allowance and alternative assets. This is strong empirical evidence of the uncertainty surrounding alternative compliance assets during the phase. Similar results obtain when cumulative historic volatility is used to predict option values. However, the true views on uncertainty about the price gap are not known; market participants could be projecting volatility using either of these methods, or any number of other schemes. The third important parameter in the

Table 11.5
Value of waiting given observed volatility.

	Cumulative volatility	Annual volatility
2008	2.04	1.53
2009	3.04	1.49
2010	2.23	1.62
2011	0.64	0
2012	0	0

trinomial model is λ, which is a scaling parameter. In this chapter λ is fixed, allowing the volatility to capture the variance in the tree.

11.5 Results and discussion

The first result is the value of waiting to exercise the spread option. These are calculated at each of the realized nodes in the trinomial tree, given the historic volatility and current price spread. The results are reported in table 11.5. Two volatility measures are compared: cumulative price volatility and a one-year history of recent volatility. If the phase-long history is taken as a measure of volatility (labeled cumulative volatility), the value of waiting is higher than the value of exercising, as seen in the right column of table 11.3. As the end of the final period approaches, the value of exercising the option disappears. If the option expires, then it is worthless.

These results must be interpreted with some care, because the option values are conditional on the parameters used to calculate them. However, under plausible parameters derived from publicly available prices and observed historic volatilities, the trinomial model indicates that the value of waiting exceeded the exercise at all nodes except the final year. At that point the option value is by definition the exercise value. This suggests that given available information, installations would recognize the value of waiting at all times during the phase, until the final period. Because the option was in the money at the conclusion of the final year in the phase, any installation with remaining capacity to capture the price spread by surrendering the alternative assets would find it financially attractive to do so.

Combining the results for the option values (table 11.5) with the exercise values (table 11.3), it appears that installations did exercise the spread option most often when the gains were largest. The final column of table 11.2 reports the percentage of allowed exchanges that took place at each node. The simple correlation between the share surrendered at each node and the exercise value is 0.97 for cumulative volatility and 0.99 for annual volatility.

Note that using the annual volatility measure, the model indicates that exercising the option is optimal at the penultimate node. The option value is calculated to be exactly equal to the price gap. An alternative interpretation is that an installation could be indifferent between exercising and not.

A second interesting result from this model is the implied volatility derived from the observed data. This exercise attempts to answer the following question: given the realized outcomes, what volatility is implied by the behavior of market participants? Another way of thinking about this is to compute the volatility that would explain an observed price step. This can be thought of as the market expected volatility that explains the observed prices on and annual basis, and it therefore relative to the annual volatilities reported in table 11.4. These results are reported in table 11.6. The implied volatilities are reported for a range of values of λ.

These results are striking in the large changes in the implied volatility over time that are derived from the observed data. This suggests that expectations about the evolution of the price spread varied widely and changed considerably over the course of the phase. In 2009 the implied volatility is very similar to the historic volatility that was observed. However, in other years the implied values are much higher. This is particularly true in 2010 and 2012. Additional information about the use of alternative compliance assets that became available in those years may have contributed to the large implied volatility. For example, the closures of the Chicago Climate Exchange in 2010 and BlueNext in 2012 may have contributed to uncertainty about the future of the price

Table 11.6
Implied volatilities.

	Observed price change	$\lambda = \sqrt{3}$	$\lambda = \sqrt{2}$	$\lambda = 1$
2008	−4.15	0.18	0.22	0.32
2009	−1.01	0.27	0.34	0.47
2010	+2.02	1.88	2.31	3.26
2011	−0.24	0.53	0.65	0.92
2012	+2.80	1.17	1.43	2.03

spread as trade had to revert to OTC markets. The development of linking directives during 2012 is a probable explanation for increased uncertainty about the future value of the spread option.

It is worth considering what drives the volatility of prices. Hintermann (2012) develops a continuous-time option approach for allowance prices based on emissions. Emissions vary as a function of economic activity and weather realizations. Another set of factors affect the market for carbon offset credits, including supply and expectations about the usefulness of buying alternative compliance assets. In combination all of these factors affect the volatility of the equilibrium outcome observed in the price series.

The results presented above offer a new explanation for the timing of surrendering alternative compliance assets to country registries during the Kyoto compliance period. Testing the hypothesis that emitters view the possible arbitrage of substituting compliance assets has not been testable until the last year of compliance for the phase was observed in early 2013. The large number of assets surrendered in the last year of the period was made possible by several factors: increased availability of ERUs, the widening price gap, and growing awareness over the years of the phase about the potentially valuable trade. The results above suggest that the value of waiting to exercise the spread option was an important factor. There are several important aspects of this problem that are not addressed in the simple model, and a few of them are discussed here.

One concern with use of non-allowance assets is that there might be fraudulent credits. This creates additional uncertainty about the alternative compliance asset. Czech and Hungarian registries were compromised in January 2011 and March 2010, respectively; these security breaches involved the recycling or theft of surrendered credits. This could be an explanation for the large implied volatility in later years of the period. The fraud problem was sufficiently severe as to warrant the temporary closure of exchanges for as long as three months in early 2011. To the extent that the uncertainty affected the price and volatility of offsets relative to allowances, this model does account for effects of fraud. A casual inspection of the data in tables 11.3 and 11.4 suggests that the market was affected less at the end of 2010 than at the end of 2011. The robustness of this observation is not investigated here.

The model assumes that emitters are focused solely on the price of the allowance and alternative assets. However, other potentially important factors are not expressly modeled in the analysis. One obvious possibility is varying degrees of risk aversion among emitters.

Another are concerns about *how* installations choose to comply. The additionality of offsetting credits has been questioned. Substitution between allowances and offsets relies on strict additionality for overall emissions to meet a target. It is possible that installations (and their parent firms) might be concerned about possible reputational effects from surrendering large numbers of offsets. Given that the political economy of caps in national allocation plans is poorly understood, and that the reputations could be closely related to windfall gains from allowance allocations, this seems like a potentially interesting area for future research. Issues of certification and changes in certification standards before Phase III have added additional uncertainty to the usage decision. The significance of those factors is not identified here.

Finally, while this model does an adequate job of explaining the timing of spread option exercise, its predictions regarding the quantity surrendered are less clear. Even though the number of installations exercising the option increased dramatically in the final year of the phase, the aggregate number surrendered did not approach the maximum number allowed by participating installations. This incomplete exercise of the option (some $\alpha < \bar{o}$) is not accounted for in the model presented here.

Acknowledgments

The author wishes to thank Luca Taschini and other participants in the 2013 CESifo Venice Summer Institute for useful comments.

Notes

1. See, for example, Fitzgerald et al. 2013 and Braun 2011.

2. Carbon offset credits take several forms. The most widely traded are certified emissions reductions (CERs), but emissions-reduction units (ERUs) are also widely used. Other types of carbon offset assets are certified for surrender to registries, though this varies between member states.

3. For the full details of existing and proposed linkages between emissions trading systems as well as linkages to various offset assets, see Ranson and Stavins 2014.

4. Even with asset-level surrender data, we cannot observe the time and price of purchase, and therefore assess the portfolio of compliance assets.

5. Factors explaining the type of national allocation plan that was drafted are interesting in themselves, but not explored here.

6. Hintermann (2012) develops a model of EU ETS allowance prices that explicitly models the important stochastic elements of electric output and switching in dispatch orders.

7. Questions of market power in both output and allowance markets are interesting but not addressed here. See Hintermann 2011.

8. Depending on the production activity, this decision may be effectively continuous so that $e^* = \int E(y_i(t))dt$. Electricity production, which is the most important industry covered by the EU ETS, is an important example.

9. With some abatement technologies, it may not be possible to separate output and abatement decisions, whereas other technologies may allow nesting of the abatement decision after output is determined. For a discussion of the ability of installations to abate and the implications that inability has for the allowance market, see Hintermann 2012.

10. Chevallier (2009) suggests that this may be a weak relationship. His paper focuses on the converse relationship: macroeconomic variability on allowance prices.

11. The determinative forces in imposing this phase-long cap structure instead of an annual cap are interesting, along with the political economy of the national allocation plans in general. One possible explanation is that the phase-long cap was viewed as more flexible, thereby allowing installations more choice and potentially lower compliance costs.

12. Gronwald and Ketterer (2012) explore the possibility of jump processes in the price of allowances, providing evidence from the first and second phases. Such jumps are an important stochastic element that could either raise or lower the value of the credit option.

13. In order to analyze installation-level behavior, only user accounts (100 accounts) were analyzed. The possibility that some installations use professional intermediaries trading on 121 accounts is not accounted for in these data. At this point I know of no way to link 121 accounts to potential clients.

14. Data can be found at www.eex.com.

15. A caveat is important here: Price information is observed on futures for December delivery, before cumulative annual emissions are known or verified. This price is used as the price at the node point, even though more information about emissions could come to light in the final days of the year. An alternative strategy would be to observe the price gap the day before allowances must be surrendered.

References

Benth, F., and P. Kettler. 2011. Dynamic copula models for the spark spread. *Quantitative Finance* 11 (3): 407–421.

Bento, A., R. Kanbur, and B. Leard. 2012. Super-additionality: A neglected force in markets for carbon offsets. Discussion paper 8952, Centre for Economic Policy Research.

Borovkova, S., F. Permana, and J. van der Weide. 2012. American basket and spread option pricing by a simple binomial tree. *Journal of Derivatives* 19 (4): 29–38.

Boyle, P. 1986. Option valuation using a three-jump process. *International Options Journal* 3 (7–12): 1.

Braun, N. 2011. Accounting for Permit Price Differentials in the European Union Emissions Trading System. MS thesis, Montana State University.

Braun, N., T. Fitzgerald, and J. Pearcy. 2014. Tradable emissions permits with offsets. In *Emissions Trading Systems as a Policy Instrument*, ed. B. Hintermann and M. Gronwald. MIT Press.

Chevallier, J. 2009. Carbon futures and macroeconomic risk factors: A view from the EU ETS. *Energy Economics* 31 (4): 614–625.

Conte, M., and M. Kotchen. 2010. Explaining the price of voluntary carbon offsets. *Climate Change Economics* 1 (2): 93–111.

Cox, J., S. Ross, and M. Rubinstein. 1979. Option pricing: A simplified approach. *Journal of Financial Economics* 7 (3): 229–263.

Fitzgerald, T., J. Pearcy, and N. Braun. 2013. Carbon offset use in the EU ETS. Working paper, Montana State University.

Gronwald, M., and J. Ketterer. What moves the European carbon market? Insights from conditional jump models. Working paper 3795, CESifo.

Herath, H., P. Kumar, and A. Amershi. 2013. Crack spread option pricing with copulas. *Journal of Economics and Finance* 37 (1): 100–121.

Hintermann, B. 2011. Market power, permit allocation and efficiency in emission permit markets. *Environmental and Resource Economics* 49 (3): 327–349.

Hintermann, B. 2012. Pricing emission permits in the absence of abatement. *Energy Economics* 34: 1329–1340.

Kamrad, B, and P Ritchken. 1991. Multinomial approximating models for options with k state variables. *Management Science* 37 (12): 1640–1652.

Li, M., S.-J. Deng, and J. Zhou. 2008. Closed-form approximations for spread option prices and Greeks. *Journal of Derivatives* 15 (3): 58–80.

Mansanet-Bataller, M., J. Chevallier, M. Hervé-Mignucci, and E. Alberola. 2011. EUA and sCER Phase II price drivers: Unveiling the reasons for the existence of the EUA-sCER spread. *Energy Policy* 39: 1056–1069.

Margrabe, W. 1978. The value of an option to exchange one asset for another. *Journal of Finance* 33 (1): 177–186.

Peters-Stanley, M., and K. Hamilton. 2012. Developing dimension: State of the voluntary carbon markets 2012. Ecosystem Marketplace.

Ranson, M., and R. Stavins. 2014. Linkage of greenhouse gas emissions trading systems: Learning from experience. National Bureau of Economic Research working paper 19824.

Sovacool, B. 2011. The policy challenges of tradable credits: A critical review of eight markets. *Energy Policy* 39 (2): 575–585.

Trotignon, R. 2012. Combining cap-and-trade with offsets: Lessons from the EU-ETS. *Climate Policy* 12 (3): 273–287.

Trotignon, R., and B. Leguet. 2009. How many CERs by 2013? Working paper 2009-5, Mission Climat.

Venkatramanan, A., and C. Alexander. 2011. Closed form approximations for spread options. *Applied Mathematical Finance* 18 (5): 447–472.

Contributors

Nathan Braun, Industrial Economics, Inc.

A. Denny Ellerman, European University Institute

Timothy Fitzgerald, Montana State University

Marc Gronwald, University of Aberdeen

Wolfgang Härdle, Humboldt-Universität zu Berlin

Peter Heindl, Centre for European Economic Research

Philipp Hieronymi, University of Illinois at Urbana-Champaign

Beat Hintermann, University of Basel

Frank Jotzo, Australian National University and Centre for European Economic Research

Andreas Lange, University of Hamburg

Stephen Lecourt, Chaire Economie du Climat and Université Paris Dauphine

Ralf Martin, Imperial College London

A. J. Mulder, University of Gronigen

Mirabelle Muûls, Imperial College London

Clement Pallière, CDC Climat Research, Institut d'études politiques de Paris

Jason Pearcy, Montana State University

Oliver Sartor, IDDRI and Université Paris 1

David Schüller, University Duisburg-Essen

Stefan Trück, Macquarie University

Ulrich J. Wagner, Universidad Carlos III de Madrid

Rafał Weron, Wrocław University of Technology

Peter _. Wood, Australian National University

Index